WEIRD & WONDERFUL WILDLIFE

Weedy Sea Dragon (*Phyllopteryx taeniolatus*). This exquisite relative of the more familiar sea-horses lives around the Australian coast. Its shape and colour enable it to disguise itself as pieces of floating seaweed. This one was photographed off Penguin Island, Tasmania.

WEIRD & WONDERFUL WILDLIFE

Michael Marten, John May, Rosemary Taylor

SECKER & WARBURG
London

First published in England 1982 by
Martin Secker & Warburg Limited
54 Poland Street, London W1V 3DF

ISBNs: Hardback: 0 436 27342 X
Paperback: 0 436 27343 8

Designed by Richard Adams

Phototypeset by Tradespools Ltd, Frome, Somerset
Printed and bound in
Hong Kong by Dai Nippon Printing Company/Imago Publishing

Also by the same authors:
Curious Facts (John May et al)
The Radiant Universe (Michael Marten, John Chesterman)
Man to Man (John Chesterman, Michael Marten)
Worlds Within Worlds (Michael Marten, John May, John Chesterman, John Trux)
An Index Of Possibilities (John Chesterman et al)

Mandrill (*Mandrillus sphinx*)

FRONT COVER. The blue and red colouring of the Mandrill's bare facial skin makes it the most spectacular of the monkeys. Its buttocks also have bare skin coloured blue and red, matching the face. When excited, the colours increase in brilliance, the chest turns blue and red spots appear on the animal's wrists and ankles. Zoologists now believe this colouring is used not just for sexual and defensive or aggressive displays, but as a kind of language to aid social interaction and promote group cohesion. Mandrills live in groups of up to 50 individuals with a strict hierarchy. Subordinate males present their brightly coloured posteriors to dominant ones to affirm their continued submission. Inhabitants of the jungles of Nigeria and the Cameroons, Mandrills are omnivorous and will eat anything from roots and fruit to worms, insects, snakes and small mammals. They grunt and chatter, shake their head and shoulders to indicate that they want to be groomed, and vigorously slap the ground with one hand when angry.

Southern Elephant Seal (*Mirounga leonina*)

BACK COVER. Flipping pebbles onto its back, this animal is described in more detail on page 81.

CONTENTS

Okapi (*Okapia johnstoni*). The 14-inch long tongue of the Okapi is primarily used to reach the higher branches of forest bushes, pull them close and strip them of their leaves. As here, it is also useful for washing the face. A relative of the giraffe, the Okapi is a horse-like animal six to seven feet long which lives in the dense tropical forests of central Africa. Its brown body with white stripes on the buttocks and legs help to camouflage it and its acute hearing enables it to avoid humans, with the result that it is rarely sighted and little is known of its behaviour in the wild. It was not discovered by western zoology until the last decade of the 19th century and not identified until 1901. Okapis lead a solitary existence, coming together only to mate. Afterwards, the female goes off alone to give birth. The photograph shows a female, since the males have a pair of short horns between their ears.

INTRODUCTION

This book is first of all a celebration of the diversity of life. Over one million different species of animals have been classified and named, and at least four times that many are thought to exist. Such variety is almost beyond comprehension, like the stars in a galaxy. But it is the complexity of each of these creatures that really beggars the imagination: the different sizes, shapes and colours, their senses and languages, their methods of defence and reproduction, their migrations and social interactions.

Almost any human quality one cares to name is mirrored in one animal or another. There are bats that conduct "maternity wards", weaver birds that build "apartment houses", and ants that milk aphids much as we milk goats or cows. Then there are the capabilities of animals that we do not share and can barely imagine – the extraordinary navigations of fish and birds, the ultrasonic communications of whales and dolphins, the abilities to regenerate lost limbs, reproduce asexually, live symbiotically with light-producing bacteria, survive in suspended animation, and change shape and colour at will.

In spite of a century of careful zoological investigation, our knowledge of this variety is tenuous, fragmentary and frequently contradictory. Only one or two thousand species have been studied in any kind of depth. In the case of the vast majority of animals, we know a little about their anatomy and physiology – from studying skeletons and carcasses – and almost nothing about their life in the wild. It is like describing humans as a result of examining a few corpses and the contents of their stomachs. One gets a sense of some of the things these strange bipedal creatures do, but their everyday behaviour remains a mystery.

It is necessary to emphasise the level of our ignorance because there is a prevailing opinion that we understand the basic principles that shape the lives of all animals. This confidence comes from the overwhelming scientific acceptance of Darwin's theory of natural selection. What Darwin said and what has been said since in Darwin's name are two very different things. The idea of "the survival of the fittest" has been used to justify European imperialism, male dominance and the exploitation of other animals by humans. The modern view is that it is an animal's ability to adapt when circumstances change to its disadvantage, rather than its ability to dominate when conditions suit it, that ensures survival.

Another common interpretation of Darwin is that "nature is red in tooth and claw". Western zoologists have emphasised the importance of competition as a factor in evolution: competition for food, mates and territory. The alternative concept of mutual aid, developed by Kropotkin not long after Darwin's *Origin Of The Species*, has been relegated to scientific oblivion.

Mutual aid is the idea that cooperation within species and between species complements and softens competition, and is ultimately a more important factor in evolution. Darwinian competition isolates every individual creature, pitting one against the other in a ceaseless struggle. Kropotkin suggested this was the exception rather than the rule.

An interesting case of mutual aid is the cleaner fish shown on page 129. These animals can be described as "living toothbrushes", since they spend much of their time cleaning out the mouths and gills of other fish. They even establish "cleaning stations" along coral reefs to which other fish come and wait their turn to be cleaned. And these other fish apparently recognise the cleaning station as neutral ground (or water) and do not molest each other, even though some of them ordinarily prey on each other.

There is a deeper criticism of the prevailing scientific view of animals – namely that it is entirely functional. An animal's shape, appearance and behaviour are described exclusively in terms of how they contribute to its survival. Does its colour help to camouflage it? Does its spectacular rump act as a sexual stimulant? The animal's life is seen as a series of strategies to acquire food, deter predators and propagate the species.

We even begin to think of ourselves in these terms – courtship rituals and submission postures, territorial defences and aggressive displays. Such ideas can be informative, but they are limited and dangerous when taken to be a complete description. They disregard the non-quantifiable aspects of existence, reducing animals to no more than a complex bundle of survival functions.

The limitations of the scientific view were explored by Barry Halstun Lopez in his remarkable book *Of Wolves And Men*. "I was made ever more uneasy," he wrote, "by how much fuller the wolf was as a creature in the mind of the modern Eskimo. If you examine what they have to say, if you watch Eskimos hunt, you discover something about wolves; but you also discover something about men and how they envision animals. For some, the animal is only an object to be quantified; it is limited, capable of being fully understood. For others, the animal is a likeness to be compared to other animals. In the end it is unfathomable."

This ultimate mystery implies respect for a creature's existence as a separate being, with its own nature that transcends our human explanations.

We live at a time when the rapid growth of human population, and the resulting requirements for land, water and fuel, are driving an increasing number of species to extinction. It is frightening how many of the animals shown in this book are endangered. It is even more frightening how much of the destruction we are causing is due to ignorance, thoughtlessness and greed.

What hope there is lies in our ability to see ourselves in a new relationship to other animals and to the biosphere as a whole. The images we have of animals, the ones we carry in our heads, are composites made up of childhood imaginings and experiences, cultural prejudices and scientific perceptions. They tell us little about the real nature of beasts, much more about our own attitudes.

It has become a truism that we have lost touch with nature. But when and how did this loss occur? Most modern authors ascribe it to the industrial revolution and the urbanisation that followed – people living in cities and towns have no direct connection with wild creatures. But the historical roots of our alienation from other animals are much older, dating back to the early days of human agriculture when the first monotheistic religions, especially Judaism and Christianity, were struggling to replace the animist cults, with their worship of animal gods and spirits. The Christian fathers taught that only humans have souls, and that God made the animals for human pleasure and use. This idea was taken to its logical conclusion by Descartes in the 17th century, when he argued that animals were really no more than complicated machines.

The importance of the theory of evolution is that it demolished this arrogant conception we had of ourselves as unique beings, quite different to all other living organisms. But the change of consciousness that Darwin initiated is very far from complete. Perhaps the most important work in this area in recent years has been the experiments in communicating with dolphins and apes. They force us to recognise that we can no longer think of other creatures as just "dumb animals".

John Lilly, who has been working on communicating with dolphins for over 20 years, suggested back in 1958 that once an animal learnt to speak our language, it would "demand equal rights with men". The idea that animals have rights – to live freely according to their nature, and not to be captured, slaughtered, exploited or experimented on by us – is at the root of the movement for animal liberation.

In his book *Animal Liberation*, Peter Singer argued that we extend basic rights to all humans not because of intelligence or humanness, but because we are all sentient beings. Science has shown that animals are also capable of suffering pain and fear and the traumas of imprisonment. Therefore the same morality that condemns oppression and exploitation of other races or classes of humans should condemn our oppression and exploitation of other species.

These ideas about animal rights, interspecies communication and the limits of the functional view of other creatures are starting points towards developing a new sense of our position in the web of life on this planet. This book is dedicated to helping to achieve that end.

A Note on Animal Names Wherever possible, we have indentified animals by both their common and their zoological name. The latter is in Latin and usually consists of two words, printed in italics. The first word indicates the animal's genus, the second its species. When there are three names, as with the Pacific Walrus (*Odobenus rosmarus divergens*), the third word indicates the sub-species.

In some cases it has been possible to identify only an animal's genus. These are written, for instance, *Aplysia* sp. or *Labroides* sp. in order to indicate that it is not known which particular species of sea hare or cleaner fish they are.

There are also cases where not even the genus of an animal is known. This occurs particularly amongst the microscopic creatures, the jellyfish and some of the insects. We have identified these animals according to the zoological family, order or class to which they belong. By convention, these names are not printed in italics in order to distinguish them from genus and species names. For example, the deepsea hatchet fish on pages 186–187 are members of the family of Sternoptychidae.

A SENSE OF PLACE

Animals do not exist separately from their habitat. They are part of the environment, are shaped by it and, in turn, help to shape it. They are as one. The hooves of the Rocky Mountain Goat are specially adapted to its cliff-side existence. The Weddell Seal is supremely fitted for life amongst and beneath the ice floes of the Antarctic. Long obscured by the sight of individual specimens in zoos or natural history museums, this fundamental unity between animal and environment is at the root of modern zoology. And it explains why, more than commercial exploitation or hunting, it is the invasion and destruction of wild habitats by humans that has brought so many species to the brink of extinction.

Andean Condor (*Vultur gryphus*)

Soaring off the Peruvian coast, this rare and majestic creature has the largest "sail area" of any living bird and divides its year between the Andes and the coast, where it feeds on the remains of dead sea lions. With its 10-foot wingspan, it can fly at 35 m.p.h. and rise to 15,000 feet, relying on thermal currents and rarely flapping its wings. It attacks and kills animals up to the size of a calf and its powerful beak will tear a hole in any carcass. Andean Condors mate and nest every two years, the female laying a single egg in a lofty cliff cave. Andean Indians annually capture two of the birds for Yawar, the celebration of blood. They are tied to the backs of two bulls, dragged round the market square in a wild ride, and then released. The ritual represents the triumph of the Indian spirit (the condor) over the Spanish conquerors (the bull).

Weddell Seal (*Leptonychotes weddelli*)

OVERLEAF. Lazing near a human settlement in the Antarctic, this beautiful animal's life is dominated by its glacial environment. Its large eyes are adapted to the twilight conditions beneath the ice and its front teeth protrude so that it can more easily gnaw through the ice in order to surface and breathe. It is the deepest diver of all seals, descending to 1,000 feet and orientating itself by echolocation. It can stay submerged for an hour or more because of its special ability to tolerate large amounts of carbon dioxide in its blood. Up to 10 feet long, its only enemy is the killer whale.

Hawksbill Turtle (*Eretmochelys imbricata*)

Shadowing a school of fish in a pure blue sea, this is the smallest of the marine turtles at under three feet long and has the unfortunate distinction, when young, of being the source of tortoiseshell. This is because its carapace consists of mottled horny plates which overlap like shingles towards the back. As a result, the species is threatened with extinction, especially the Atlantic population which assembles every three years to nest on the coast of northern Colombia. Like other marine turtles, the Hawksbill uses its forelegs like fins, to propel itself through the water, and its hindlegs as rudders for steering. It cannot retract its head or legs beneath its shell, and it feeds on fish as well as seaweed and algae.

Spotted Cuscus (*Phalanger maculatus*)

OVERLEAF. This nocturnal, tree-dwelling marsupial lives in New Guinea in the ecological niche normally occupied by monkeys. It can be as small as a rat or as big as a large domestic cat, with a tail as long again as its body. Its thick fur coat comes in a variety of colours depending on its age, health and locality. It eats both leaves and small animals. Cuscus species resemble the sloths of Central and South America in being very slow-moving, with low body temperature and low metabolic rate. This one's coat is spotted with dirt.

Sand Snake (*Psammophis schokari*)

This slender snake moves with great rapidity in the deserts of Africa and the Middle East. It has been described as giving the impression of "being shot like an arrow from a bow". Its diet includes such desert creatures as lizards, small rodents and other snakes. Its venom is harmless to humans but potent against its prey.

Galapagos Land Iguana (*Conolophus subcristatus*)

OVERLEAF. When Darwin visited James Island in the Galapagos in 1835, he wrote of these animals: "We could not for some time find a spot free from their burrows on which to pitch our single tent." Today the species is almost extinct, primarily because the vegetation on which it feeds has been stripped by goats introduced by colonists. It is also preyed on by the Galapagos Hawk and shot by humans, who regard it as a delicacy. A foot and a half long, its diet includes cactus fruits, the spines of which pass without adverse effect through its digestive system and are excreted. This individual is seen perched on a crater rim on Fernandina Island.

Scrawled File Fish (*Aluterus scriptus*)

Up to four feet long, this unusual fish is named for the texture of its skin, which is set with tiny scales so abrasive that it was once used as sandpaper. Its Australian common name is the "leather-jacket". It lives in seaweed fields and eats sea grass, sea anemones and barnacles which it removes with its incisor teeth and then crushes with teeth in its throat. When threatened, it stands vertically amongst the seaweed and imitates its wavy motion. On this occasion it has adopted the same defence posture, although caught in the open above the remains of a wreck.

Rocky Mountain Goat (*Oreamnos americanus*)

OVERLEAF. The few remaining herds of these bearded, pure white goats live 13,000 feet up, above the timber line, in the American and Canadian Rockies. Once widespread, they were hunted close to extinction in the 19th century. Their hooves have concave undersides with sharp edges that dig into the rock, enabling them to adhere like suction cups on the steepest cliff faces. They feed on moss, lichen, grass and leaves. With few predators apart from humans, their main problem is the saturation of their delicate wool when it rains, which can lead to pneumonia and death. Adult males stand 40 inches at the shoulder, weigh 300 lbs and are agile and powerful animals.

Goggle-Eye (*Priacanthus hamrur*)

Some 16 inches long, the Goggle-Eye or Big Eyes is common in the Red Sea and Indian and Pacific Oceans. It is mainly active at night, hiding in crevices and caves during daylight. It is not a popular eating fish except in Hawaii, where the native inhabitants call it *alalaua*, and in Thailand, where it is dried and salted.

Grey-Headed Albatross (*Diomedea chrysostoma*)

OVERLEAF. Nesting on a clifftop by the southern ocean, this is one of the smaller species of albatross, though it has a wingspan of some eight feet. The nest is a simple heap of earth and vegetable matter. The female lays a single white egg, but she does not recognise her own offspring and will feed any young bird placed in the nest. Albatrosses have adapted to an aerial existence over the ocean more completely than any other birds, gliding for weeks at a time above the cold seas of the Antarctic subpolar zone, known as the "albatross latitudes". (See also page 37.)

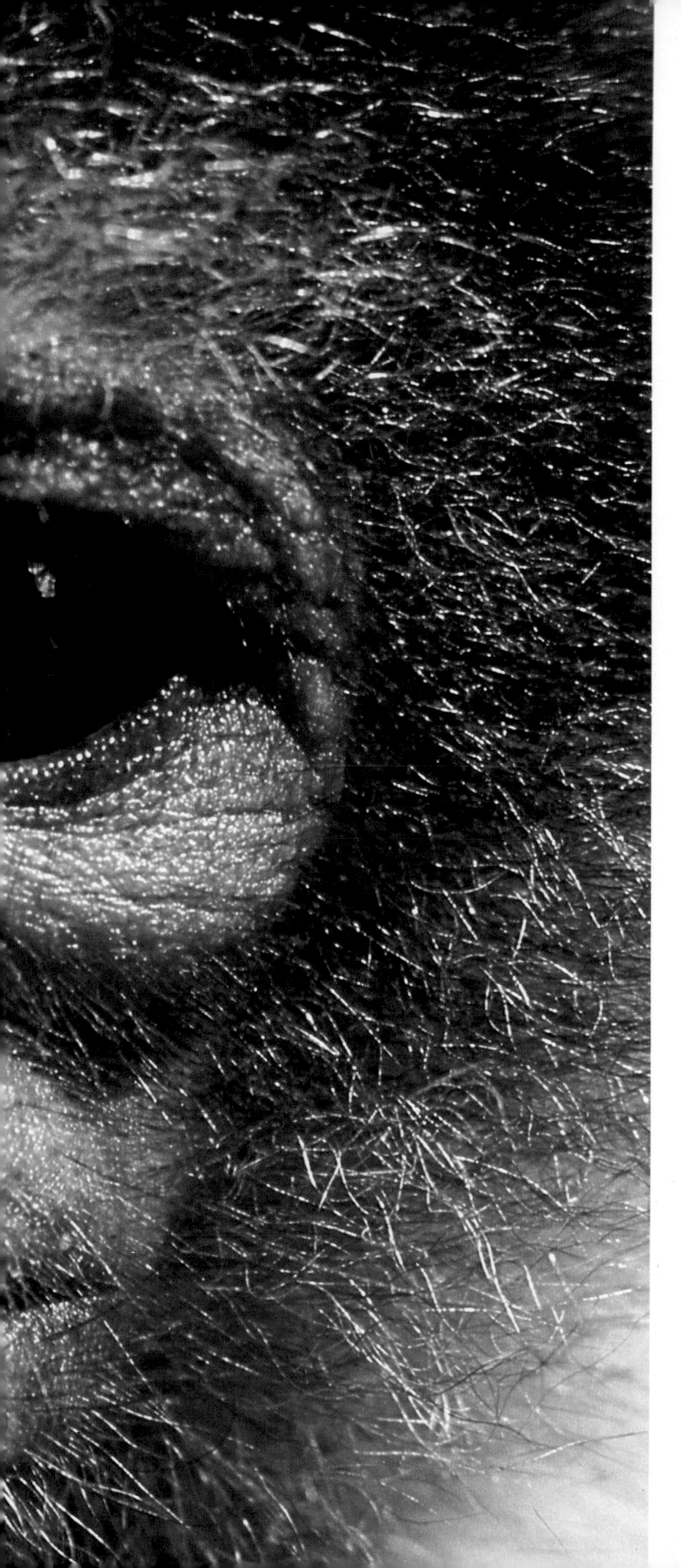

TWO

HEADS

When Darwin proposed that we are descended from the apes and, eventually, from even more primitive animals, the shock and fury of the public reaction reflected the assault on our pride. Today the theory of evolution is overwhelmingly accepted, but we still find it difficult to concede that we are really animals like others. These startling portraits reveal in sharp focus the similarities and differences between other animals and ourselves. From the uncannily human physiognomy of the chimpanzee, our closest relative, to the alien, ghost-like, almost mocking "face" of a ray, we gaze into reflections and distortions of our own features.

African Chimpanzee (*Pan troglodytes*)

Chimpanzees spend most of their lives on the ground, climbing trees only in order to pick fruit, build nests and sleep. They walk or run on all fours, supporting their weight on the knuckles of their hands. When standing erect, a full-grown male is up to five and a half feet tall. This species lives in West Africa and is primarily vegetarian, though it occasionally eats other mammals, even small monkeys. It also feeds on insects, using sticks to poke and probe termite mounds. Agile and intelligent, it is individualistic and does not form permanent groups.

Philippine Eagle (*Pithecophaga jefferyi*)

OPPOSITE. Only discovered in 1894, there are thought to be less than 100 individuals of this species left in the dense mountain forests of some of the Philippine islands. A fearsome creature standing three feet tall and weighing 9 lbs, it has huge talons and a vicious hooked beak. It attacks small domestic animals and other birds like hornbills, but its main prey is macaque monkeys. It was known as the Monkey-Eating Eagle until President Marcos of the Philippines changed its name because the reference to its dietary habits "denigrates the quality of this bird, in whose rarity and confident bearing the Philippines can take pride".

King Penguin (*Aptenodytes patagonica*)

ABOVE. Most colourful of penguins, the Kings, which are second in size to the Emperors, breed in sub-Antarctic latitudes. Like other penguins, they will eat snow and drink both salt and fresh water. Their wings have been transformed into flippers and they move through the water at speeds of up to 23 m.p.h. They feed on plankton, small fish, crabs and squid. No birds have a body temperature as low as a human's, but the King Penguins come closest with an average of 100°F (37.7°C).

Ray (Rajoidei)

ABOVE. This ghost-like face is the underside of the head of an unidentified species of ray, possibly one of the skates. Rays have almost completely flattened bodies, with their eyes on the upper half and their mouths underneath. The two eye-like features above the mouth are in fact the creature's nasal openings, while the five pairs of gill slits that distinguish rays can be seen below. The rays are found in all seas, from coastal waters to depths of 10,000 feet.

Atlantic Walrus (*Odobenus rosmarus*)

OPPOSITE. This is a young walrus, since it does not have the tusks that distinguish adult cows and bulls. It will grow into a huge animal, measuring 11 feet and weighing up to one and a half tons in the case of a male. Atlantic Walruses are found in two populations, one between Greenland and Canada, the other off northern Siberia. They live mainly on molluscs, which they prise from the seabed or from rocks with their tusks, removing the shells and swallowing the soft parts. Young walruses take time to learn the technique and are suckled by their mothers for one year. Adults will also attack seals, hacking them with their tusks and feasting on their fat. (See also page 205.)

Parrot Fish (Scaridae)

OPPOSITE, TOP. The 80 species of parrot fish live in tropical and subtropical seas along the steep ridges of coral reefs. Their teeth have fused together to form a beak-like jaw, hence their name. They have another set of grinding teeth in the throat. Up to six feet long, they are herbivorous, scraping algae and coral from the reefs, which they are largely responsible for eroding. The members of some species always swim to the same site after feeding and excrete the indigestible parts of the coral so as to form small hillocks of their droppings. Parrot fish are also unusual in secreting every night a shiny substance which forms a protective sleeping cocoon around them; they emerge from these "pyjamas" in the morning.

Grey Angelfish (*Pomacanthus arcuatus*)

OPPOSITE, BOTTOM. Another inhabitant of coral reefs, notably in the Caribbean, its scales give its skin an extraordinary lattice-like appearance. It hides in crevices in the coral at night, emerging during the day to feed on algae, worms, polyps and crustaceans. It is solitary or lives in pairs and grows up to two feet long. Juveniles of this species look quite different, being black with yellow bars across the head and body.

Grey-Headed Albatross (*Diomedea chrysostoma*)

ABOVE. More like a painting than a photograph, this is a close-up of the albatross seen nesting on pages 28–29. Amongst their other unusual characteristics, albatrosses produce in their stomachs a yellow fluid which condenses to the consistency of wax at low temperatures. High in nutritional value, it is regurgitated through the nostrils for feeding the young or for use as extra waterproofing of the feathers. It can also be shot from the bill at some velocity to deter predators. Its rancid smell is thought to give albatrosses their unpleasant musky odour.

Californian Condor (*Gymnogyps californianus*)

The bare, fleshy head of this bird is typical of members of the vulture family. Huge and ungainly, with a 9-foot wingspan, this species may soon be extinct. At the end of the last century it was open season on condors: they were easily shot and their eggs fetched £150 apiece on the collectors' market. More recently, the use of pesticides has caused them to produce thinner eggshells, increasing infant mortality. Modern farming has meant fewer dead cattle lying around for them to feed on. And the spread of towns and cities along America's west coast has driven them back into the mountains.

Pig-Tailed Macaque (*Macaca nemestrina*)

With its long muzzle and sharp nose, this intelligent monkey is amongst the largest of the macaque family, growing over two feet long. It lives in large groups, spends much of its time on the ground and has been known to overrun towns in search of food. Found throughout most of south-east Asia, the young and females are sometimes trained as helpers during the coconut harvest, but the males are too intractable and dangerous. The species has a friendship greeting unique amongst monkeys. They jerk their head up, raise their eyebrows, squint over their nose and push their lips forward in an exaggerated "pout".

Fennec Fox (*Fennecus zerda*)

ABOVE. The Fennec is common in the Arabian desert, the Sinai Peninsula and the Empty Quarter of the north-west Sahara. It lives entirely independently of water, obtaining the moisture it needs from the insects, lizards, rodents, birds and plants it eats. A nocturnal animal, it avoids the fierce heat of the day by retreating to its deep den in the sand. Fennecs grow only a foot long and are the smallest wild members of the dog family in the world. Their exaggerated ears, however, are up to six inches tall and serve a dual function: as a means of detecting prey at night, and as a highly efficient device for losing heat because of their network of blood vessels spread over a large surface area. Fennecs live in groups of about 10 individuals, and their dens are often connected in the sand.

Alpaca (*Lama guanicoë pacos*)

OPPOSITE. Related to the llama, guanaco and vicuna, the Alpaca has been domesticated for some 2,000 years and no longer survives as a wild animal. The Incas considered it a gift of the gods and devoted much attention to its selective breeding. It is valued for its wool which is light, warm and proof against rain and snow. Alpacas have a mind of their own, however, and are reputedly troublesome to shear.

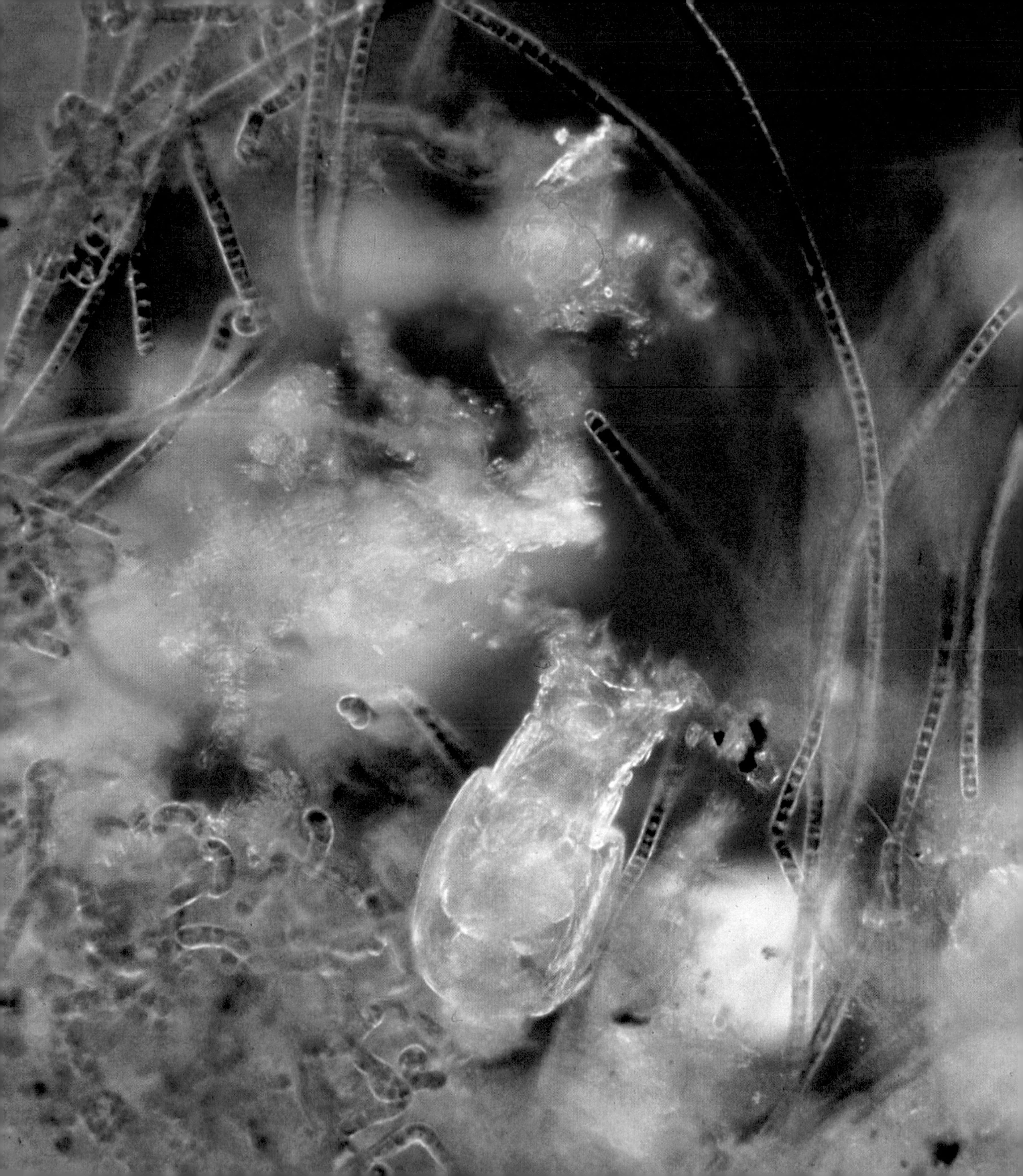

THREE

MICROWORLD

Many of the most extraordinary and beautiful creatures on Earth are so minute that we can see them only with the aid of a microscope. The billions of inhabitants of the microworld include both unicelullar animals, the Protozoa, which consist of a single cell of highly organised protoplasm, and multicellular creatures such as the Rotifers opposite and overleaf. They live in an environment that is in many respects a scaled-down version of our own: they swim in ponds that consist of drops of water or oceans that we think of as ponds; they move amongst towering forests of microscopic plant material; there are herbivores and carnivores and strategies for defence and reproduction as intricate as anything in our larger world.

Rotifers (Rotatoria)

OPPOSITE & OVERLEAF. The first explorers of the microworld dubbed them "wheel animalcules" because the regular beating of the hair-like *cilia* around the edge of their funnel-shaped mouths gives a distinct impression of spinning wheels. "O, that one could ever depict so wonderful a motion!" exclaimed an artist employed by the great 17th century Dutch microscopist van Leeuwenhoek. The cilia produce a whirling current that brings food particles to the mouth. There are over 2,000 rotifer species spread across the globe from pools of thawed water in the Antarctic to every back-garden pond. Some species have taken to the sea and others to the soil where, according to one estimate, the top two inches of a square yard of cultivated land or grass may contain up to 300,000 of the creatures. Most rotifers are females. A great many species have dispensed with males altogether, reproducing entirely by virgin birth. Others alternate between sexual and asexual reproduction, but the males they do produce are more or less degenerate: they usually live just a few hours or days, cannot feed, and are equipped only for seeking out and copulating with females. Rotifers are also distinguished by their extreme tenacity to life. When conditions become unfavourable, they become desiccated and form "cysts". In this state of suspended animation they can survive extremes of temperature, or drought, for years on end. When a thaw sets in, or they are moistened with a drop of water, they return rapidly to life, wheels spinning.

1. *Philodina* sp. A common pond-dweller, it is seen opposite feeding amongst a tangle of organic matter and filamentous algae. Magnified 240 times (×240).

2 3

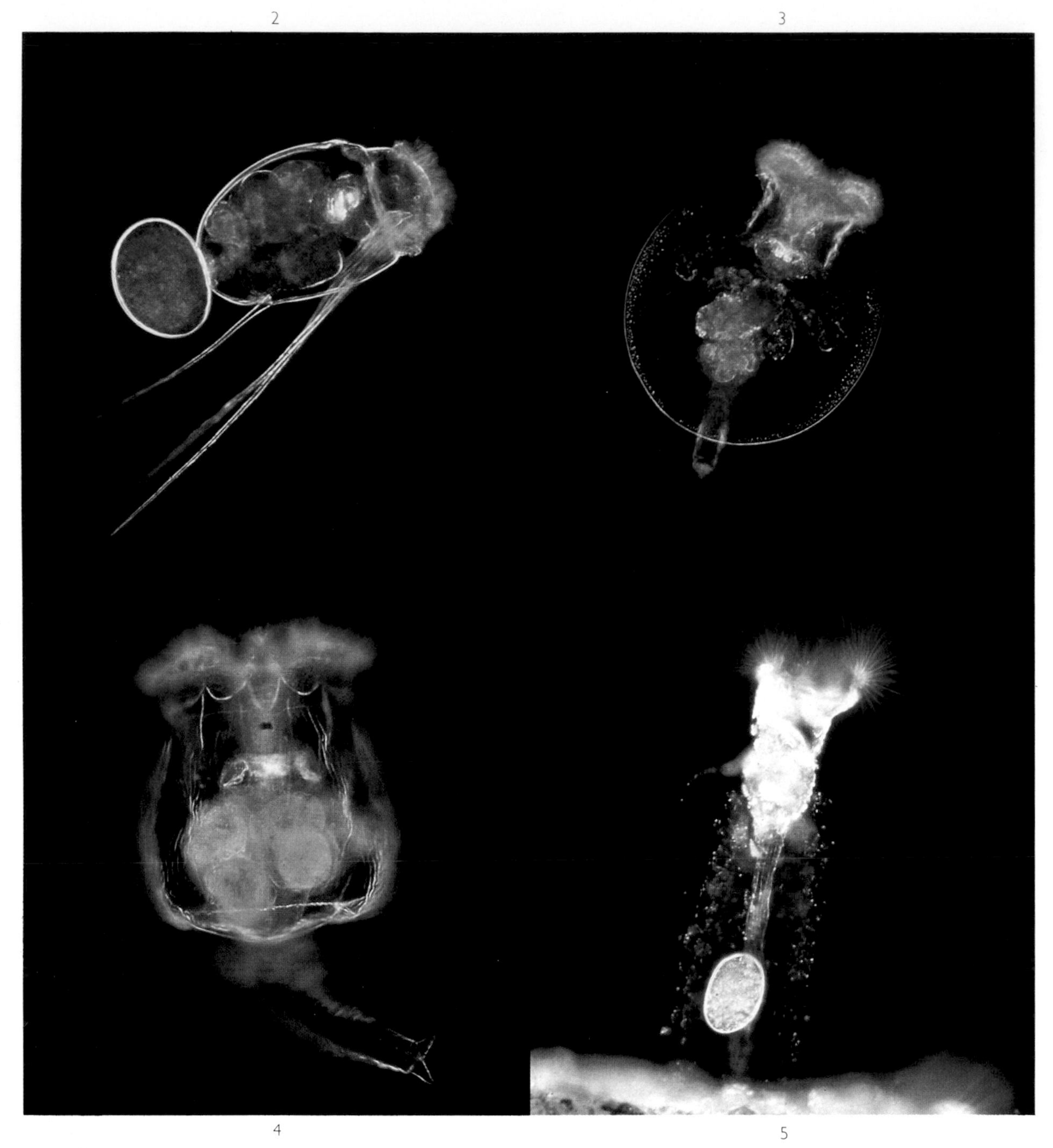

4 5

2. *Filinia* sp. A small planktonic creature, it uses its three appendages to suddenly flip itself backwards if disturbed. It is seen with its corona of cilia extended. The red spot is its eye and the large oval at the end of its body is an egg. ×375.

3. *Testudinella* sp. Usually found amongst weeds and algae at the edge of ponds, it is named for its tortoise-like habit of withdrawing its corona and "tail" into its transparent shell when disturbed. ×200.

4. *Brachionus* sp. The single red eye can be seen beneath the corona of this common rotifer. Immediately beneath the eye are its hard, pointed "jaws" or *trophi*. The two yellow shapes are developing eggs. The trunk-like foot ends in two toes which secrete a cement that anchors the animal when it is feeding. ×100.

5. *Colletheca* sp. It is a sedentary creature which uses its long, stiff cilia to enmesh prey organisms. If disturbed, it retreats into the transparent tube of mucus with which it

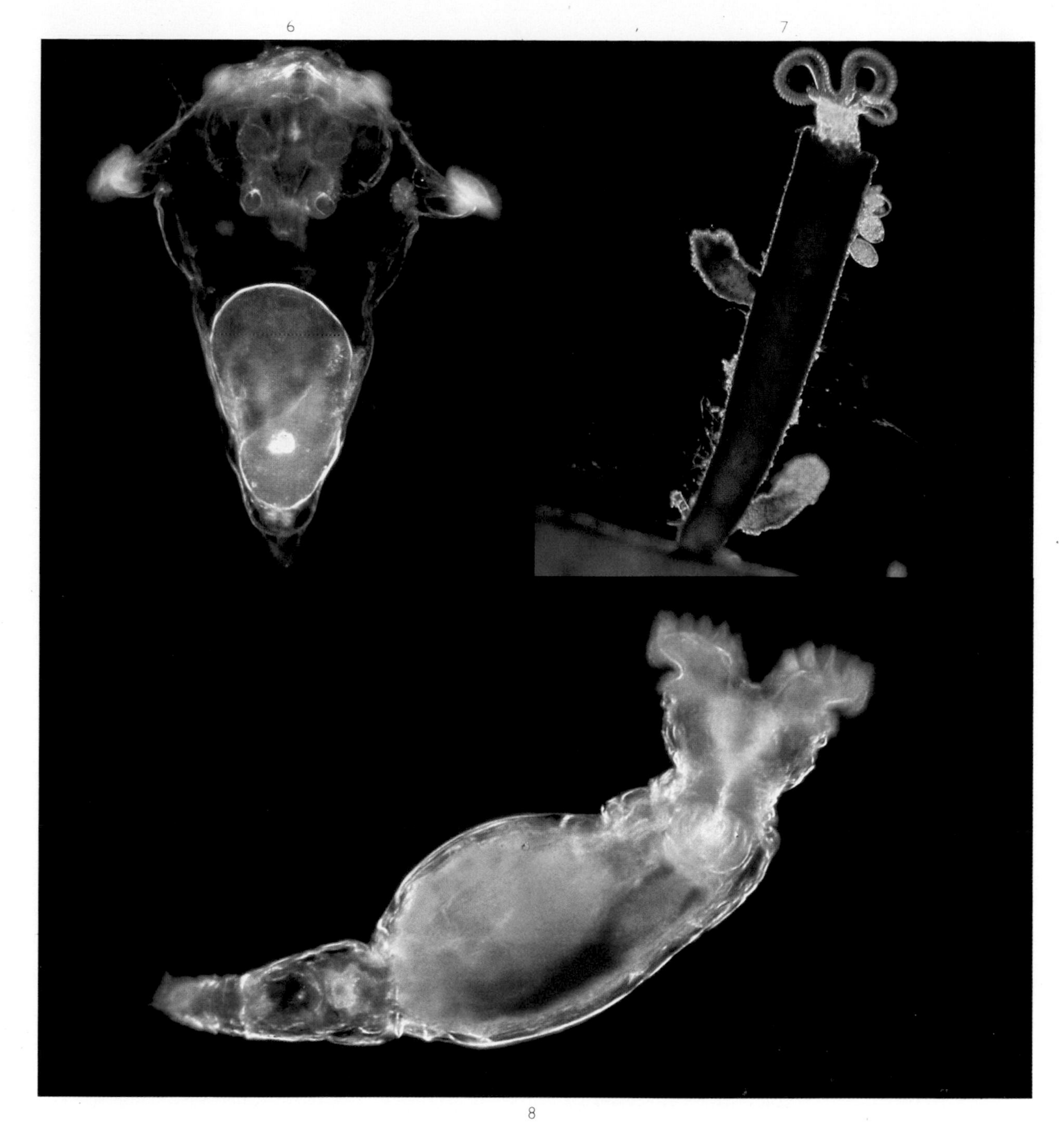

attaches itself to an object – in this case a submerged leaf. A single egg can be seen developing in the tube. ×140.

6. *Synchaeta* sp. This planktonic rotifer is distinguished by the two extra tufts of cilia on "outriggers" at either side of its body. They are an aid to locomotion. ×280.

7. *Floscularia* sp. A relatively large rotifer, it constructs a gently tapering tube from pellets of compacted debris. This often provides a base, as here, for the tubes and eggs of smaller rotifers. *Floscularia*'s ornate corona can be seen extending from the top of its tube. ×64.

8. *Philodina gregaria*. Indigenous to the Antarctic, it occurs in such large numbers that it can colour the floor of a lake rusty red. It feeds on algae, gives birth to live young, and can survive drying and freezing for indefinite periods. ×340.

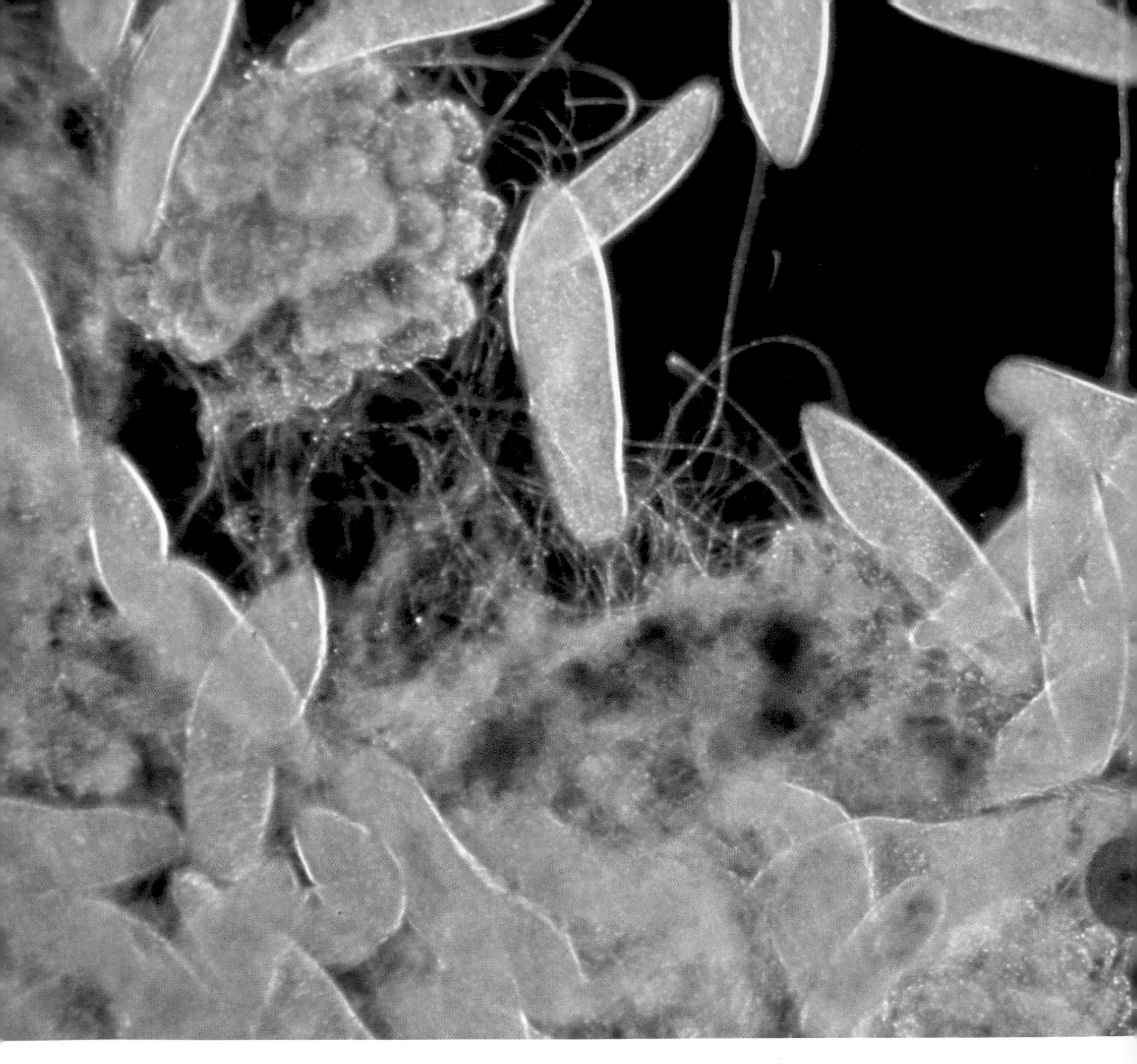

***Paramecium* sp.**

These slipper-shaped unicellular organisms, here magnified 340 times, are amongst the most common inhabitants of the microworld. Seen under the microscope, they swim back and forth under the power of the cilia which completely cover their bodies. Their movements are rapid and purposeful; when they meet an obstruction, they reverse the beat of their cilia to back away, then try a new tack. In this picture they are seen feeding on a mass of filamentous bacteria. *Paramecia* multiply by dividing lengthwise. They also conjugate, two *Paramecia* coming together side by side so that their bodies fuse; conjugation involves the transfer of genetic material, but does not directly lead to the production of offspring.

Zoothamnium sp.

ABOVE. These ciliate protozoans are found in branching colonies (left) about a twentieth of an inch across. When disturbed, each member of the colony suddenly contracts its own stalk (right), although the colony's trunk remains extended. The contraction occurs so rapidly that the eye cannot follow it. *Zoothamnium* and similar species play an important role in removing bacteria and suspended particles from water and are found in vast numbers in sewage processing plants. ×85.

Stentor sp.

OPPOSITE. This exquisite creature is one of the largest protozoa, growing about a twentieth of an inch tall when fully extended. Its name is taken from the loud-voiced hero of the Trojan Wars and refers to its trumpet-like form. Its body is covered with longitudinal rows of fine cilia and the rim of the trumpet is fringed with appendages of fused cilia, known as *cirri*, with which the animal creates a current to bring food organisms such as algae and other protozoa into its funnel. Most *Stentor* species live in still and often polluted water, but some are parasites, including one that is commonly found in the intestines of pigs and occasionally of chimpanzees and humans, where it can cause ulcers. ×550.

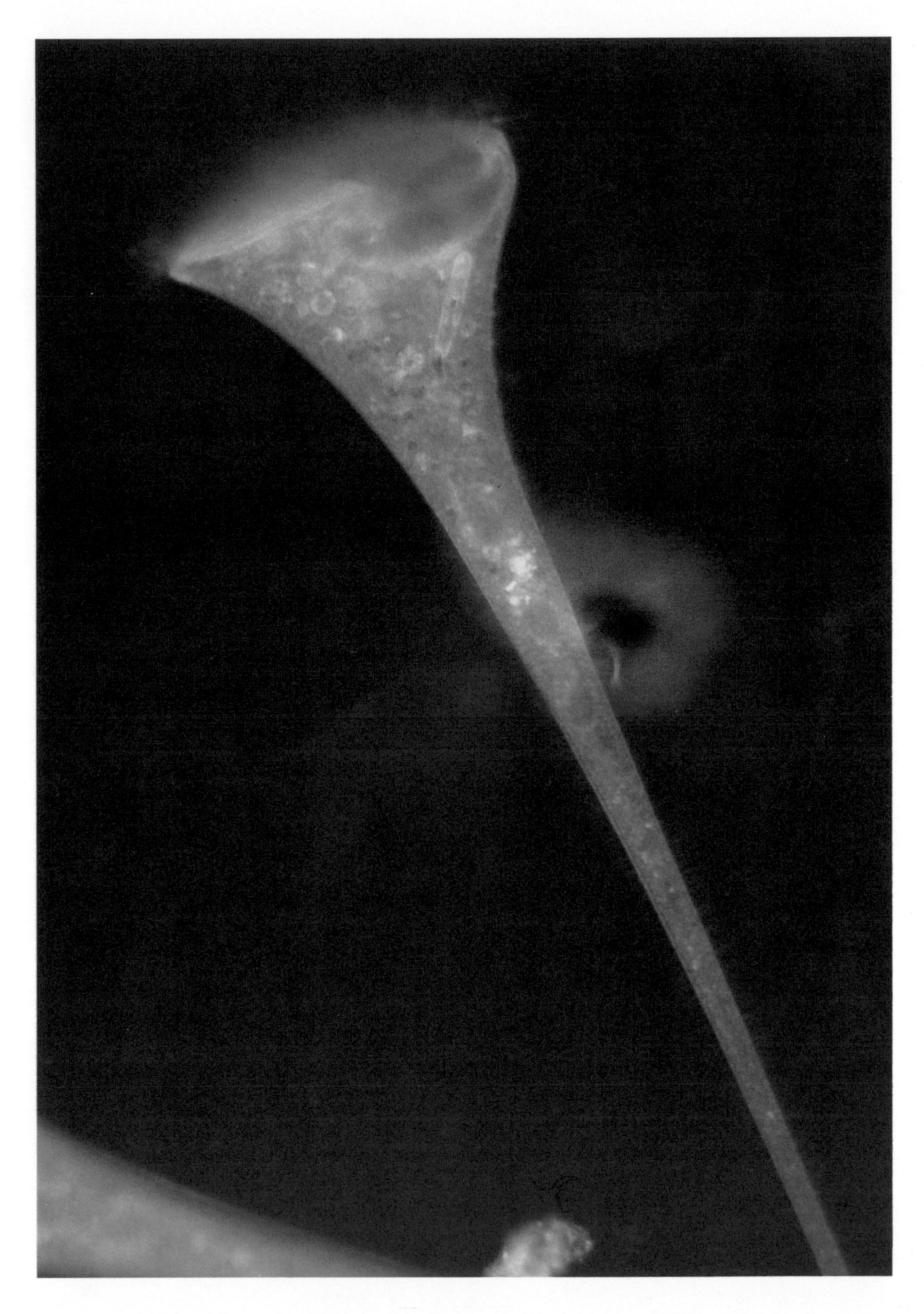

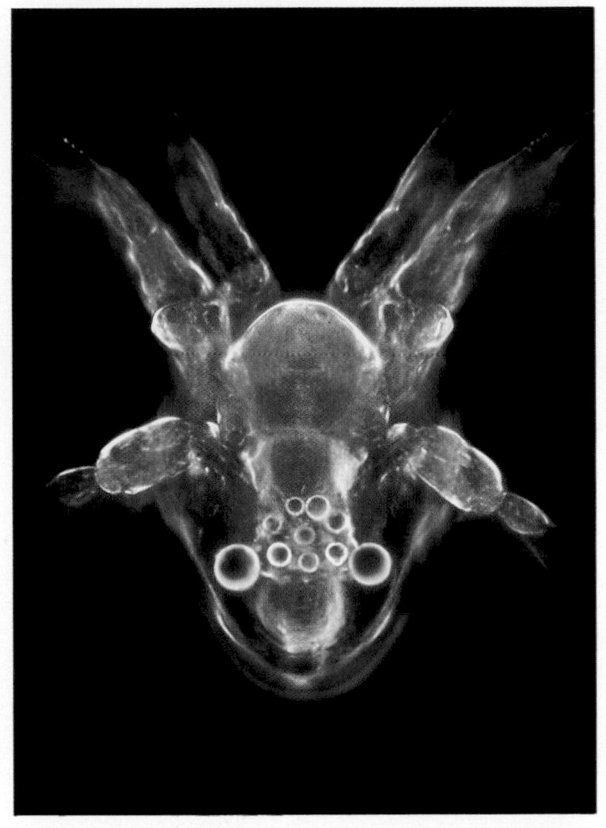

Amoeba (Amoebina)

OPPOSITE. The jewel-like appearance of this unidentified amoeba is the result of food particles it has ingested. The larger yellow form in its body is a diatom it has eaten. The animal is moving towards a mass of organic debris at the bottom of the picture. Amoebae are protozoa which often have no definite shape since they are always oozing in one direction or another. Many species alternate between living in colonies of their fellows and adopting a solitary existence. Some are parasites, such as the creature responsible for amoebic dysentery in humans. ×1,100.

***Cyclops* sp.**

ABOVE. These two bug-like creatures are larvae of the freshwater crustacean *Cyclops*, numerous species of which exist wherever there are stagnant ponds. The red dot at the front of the one on the left is its eye-spot. The yellow circles in the body of the one on the right are fat globules. *Cyclops* belong to the group of crustaceans known as Copepods, which are found in both freshwater and marine environments and are believed to be the most numerous animals on Earth after the protozoa, outnumbering even the insects. They are the main converters of plant material into animal protein and therefore play a vital role at the base of the food chain upon which all animal life depends. Magnifications are ×275 (left) and ×390 (right).

***Hydrachna* sp.**

OVERLEAF. There are over 2,400 species of freshwater mites and many of them are brightly coloured like this red-bodied, orange-legged creature. It is seen propelling itself with the swimming hairs on its legs through a tangle of filamentous green algae. Like their terrestrial and often parasitic brethren, the freshwater mites are Arachnids, the category of animals that includes spiders and scorpions. ×60.

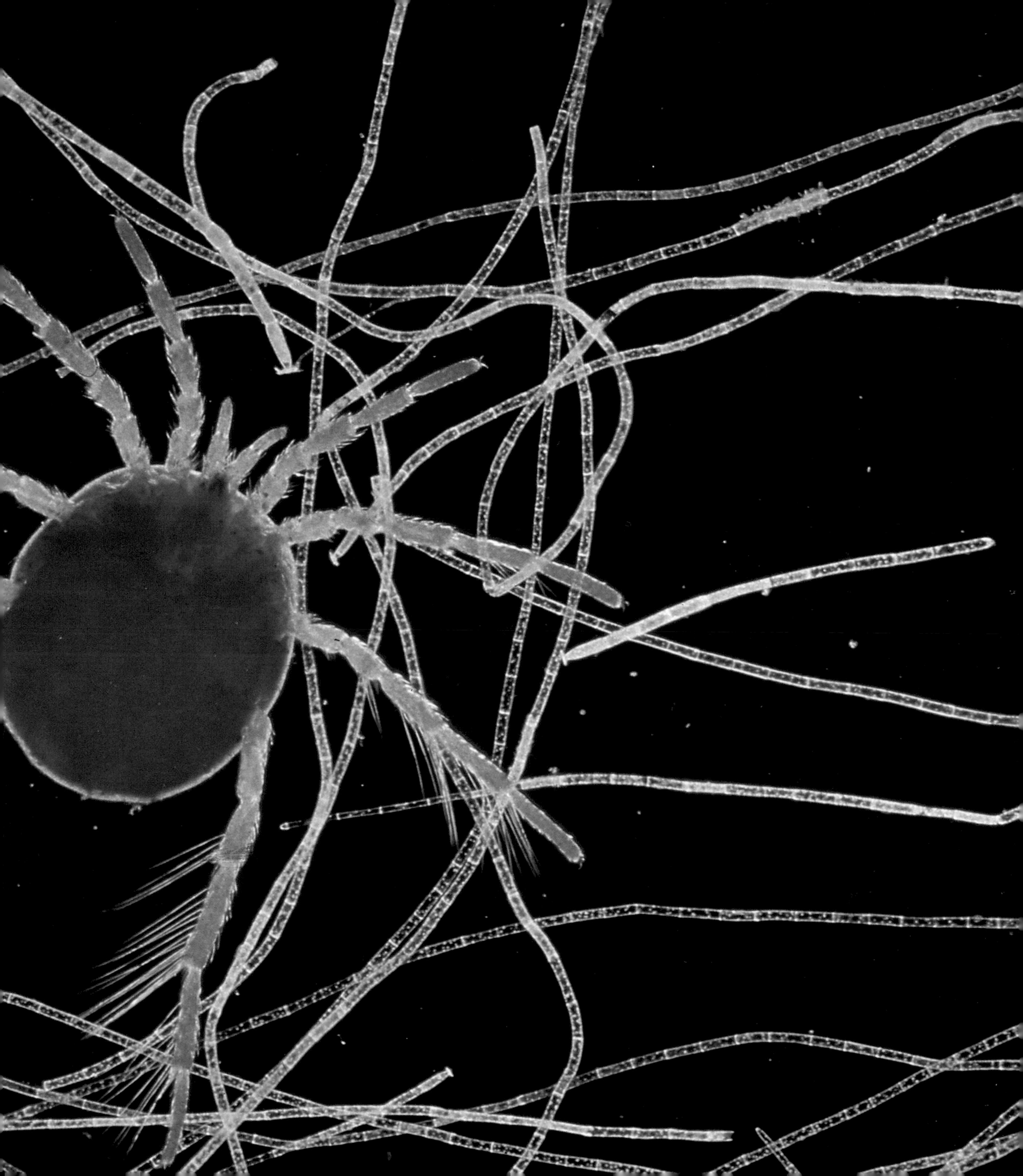

FOUR
GALLERY OF THE GROTESQUE

These strange and hideous creatures are not the product of a nightmare or a special effects laboratory. They may appear repulsive to human sensibilities, but their grotesqueness is often an adaptation to their particular ecological niche. Plaice, for example, are fish with a remarkable lifestyle. The egg hatches out to produce a free-swimming larva attached to a yolk-sac. Later, when the larva metamorphoses into the fish, its skull develops at different rates on each side. This causes its mouth to become distorted, while its left eye migrates across the head until it is on the right side of the body. This side then becomes the new top half of the fish. Its whole body flattens out and it descends to rest almost invisibly on the seabed. This extraordinary physical transformation is an adaptation to its bottom-dwelling life and is common to all members of the flounder order of fishes, including soles, turbots and halibuts. In some species the left eye migrates to the right, in others the right eye migrates to the left.

Chameleon Plaice (*Pleuronectes platessa*)

Boss-eyed and vertical-lipped, it is called the Chameleon Plaice because the skin of its top half contains red, white and black pigment cells which it can expand or contract to change colour to suit its surroundings. It lives along the coasts of western Europe, feeding on worms, crustaceans and small mussels whose shells it cracks open with powerful throat teeth. Before it began to be intensively fished for our tables, it grew over three feet long. Today a length of two and a half feet is a rarity.

Brown-Headed Tamarin (*Saguinas fuscicollis*)

Caught in mid-bite, with an insect's head in its mouth and the rest of the body still gripped in its claws, this inhabitant of the jungles of Guyana and the Lower Amazon has a hairless face, exceptionally large ears and, despite its name, predominantly black fur. Along with marmosets, tamarins are the smallest monkeys with an average body length of only some 10 inches and tails just over a foot long.

Naked Mole Rat (*Heterocephalus glaber*)

This unique hairless rodent lives an entirely subterranean life in eastern Africa south of the Sahara. It is blind, with eyes no larger than pinpricks. It has slender toes with long claws for digging in loose soil around plant roots and prominent incisor teeth for chewing the roots themselves. Its presence can be detected by the miniature craters formed on the surface where it excavates its burrows.

Vampire Bat (*Desmodus rotundus*)

After dark, the Vampire Bat flies out from its cave in search of warm-blooded victims: cattle, horses, pigs, and humans. The soft skinpads on its feet and wrists allow it to alight so gently that the victim does not awake. Its razor-sharp teeth shave off a tiny area of hair and skin, exposing a blood vessel. Its saliva contains an anti-coagulant so that it can continue licking up the blood for many minutes, until it swells visibly and can move only clumsily. It then retires to its cave for several days to digest its meal. Vampire Bats are found throughout South America.

Stonefish (*Synanceia verrucosa*)

ABOVE. This is the most venomous fish in the world. It has 16 well-camouflaged spines connected to two venom sacs. Human victims suffer 12 hours of extreme pain and, if they do not receive the anti-toxin, their body slowly grows numb and they die. Also known as the Devilfish, it grows a foot long and is found amongst coral reefs in the Indian and Pacific Oceans. The photographer entitled this picture "I hate mornings".

Giant Spiny Stick Insect (*Extatosoma tiaratum*)

OPPOSITE. Also known as Macleay's Spectre, it lives in Australia and New Guinea. Like other stick insects, it is perfectly camouflaged when sitting motionless in dry vegetation. It can also change colour, break off a limb to distract a predator and grow a replacement, and the female can give birth by herself, without benefit of male fertilisation. Note the teeth-like mouth parts and the pale green compound eyes.

Shorthorn Sculpin (*Myxocephalus scorpius*)

A bottom-dwelling, omnivorous fish, it is said to growl if held in the hand. The male grows up to 10 inches long, considerably larger than the female, and guards the eggs until they hatch. Its diet includes worms, crustaceans, and other fish, and it annoys coastal fishermen by going after their bait. This specimen was found lurking in Hodgkin's Cove, Massachusetts.

FIVE
ORCHIDS OF THE OCEANS

Nudibranchs, whose name means "naked gills", are carnivorous, hermaphrodite sea-slugs which are found in all marine environments, from the Antarctic to tropical mangrove swamps. More than 3,000 species of these creatures are known, and they come in such an extraordinary variety of colours and shapes that they are often described as the "orchids of the oceans". Nudibranchs range in size from a species which can crawl between the grains of sand on a beach to monsters weighing over 3 lbs. Related both to shellfish and the common garden snail, they have abandoned their shells to go naked into the world. As a result, they have evolved a whole array of biological and chemical defences to ensure their survival. Some species, for instance, camouflage themselves by turning the colour of the sponges they eat, and even lay eggs that colour. Others have developed the ability to swallow the sting-cells or "nematocysts" of sea-anemones and poisonous hydroids and turn them to their own use. Just how this trick is achieved is still uncertain, since nematocysts are designed to explode on contact. Nonetheless, nudibranchs like the Mexican Dancers on this page can swallow these sting-cells and pass them undigested through their intestines so that they reach special sacs in the nudibranch's own skin. In effect, these creatures have hijacked the sea-anemone's own weapons.

Mexican Dancer (*Coryphella iodinea*)

This mating pair was photographed off the western coast of Mexico. The sting-cells they "steal" from hydroids are located at the bright orange tips of their finger-like projections or *cerata*.

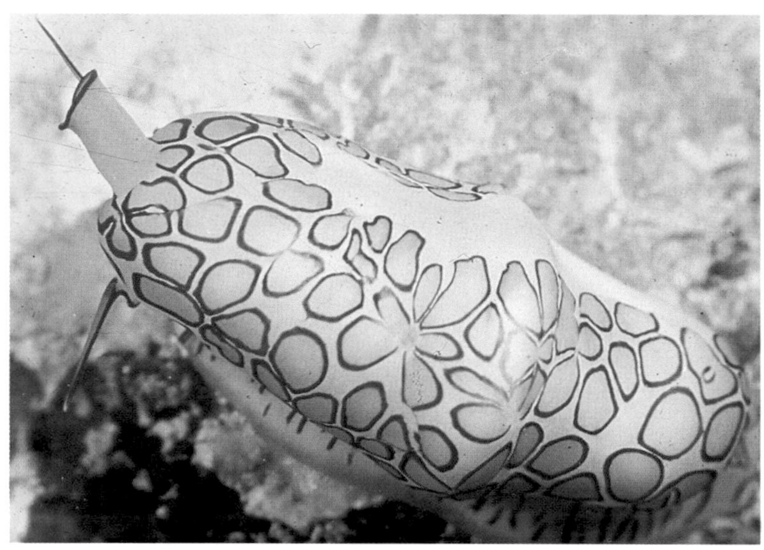

Haitian Jewel Snail (*Micromelo undata*)

TOP LEFT. One of the most spectacular inhabitants of Caribbean coral reefs, it has brightly coloured hard and soft parts and is immortalised on a Haitian postage stamp.

Casella atromarginata

TOP RIGHT. In spite of its delicate, black-and-white appearance, *Casella* is as tough as a leather boot. A tropical inhabitant of the Indian and Pacific Oceans, it feeds on encrusting sponges from Australia to Aqaba.

Flamingo-Tongue Snail (*Cyphoma gibbosa*)

BOTTOM LEFT. Like an exquisitely-patterned piece of porcelain, this species feeds on soft coral in Caribbean lagoons. It grows to just over an inch long.

***Nembrotha* sp.**

BOTTOM RIGHT. Very little is known about these nudibranchs. They are generally small and are believed to feed on sedentary encrustations like sponges. This one was photographed off Heron Island, on Australia's Great Barrier Reef.

Phylliroe bucephala

So transparent are these nudibranchs that all their internal organs are visible. They have flattened bodies and move through the water in a snake-like fashion. In the juvenile form, this species lives as a parasite on a kind of medusa (*Zanklea costata*) which is sucked dry until it is reduced to a mere appendage hanging from the nudibranch's vestigial foot.

White Sea-Slug (*Dirona albolineata*)

This species lives in the cold waters of the Japan Current where it touches the north-western coast of America. Harmless themselves, White Sea-Slugs gain protection from their resemblance to one of the stinging nudibranchs known as "nettle-slugs". The individual on the right is carrying the coiled white egg sacs of an internal parasite; once attacked by this worm, the nudibranch can live for several weeks but will never reach sexual maturity.

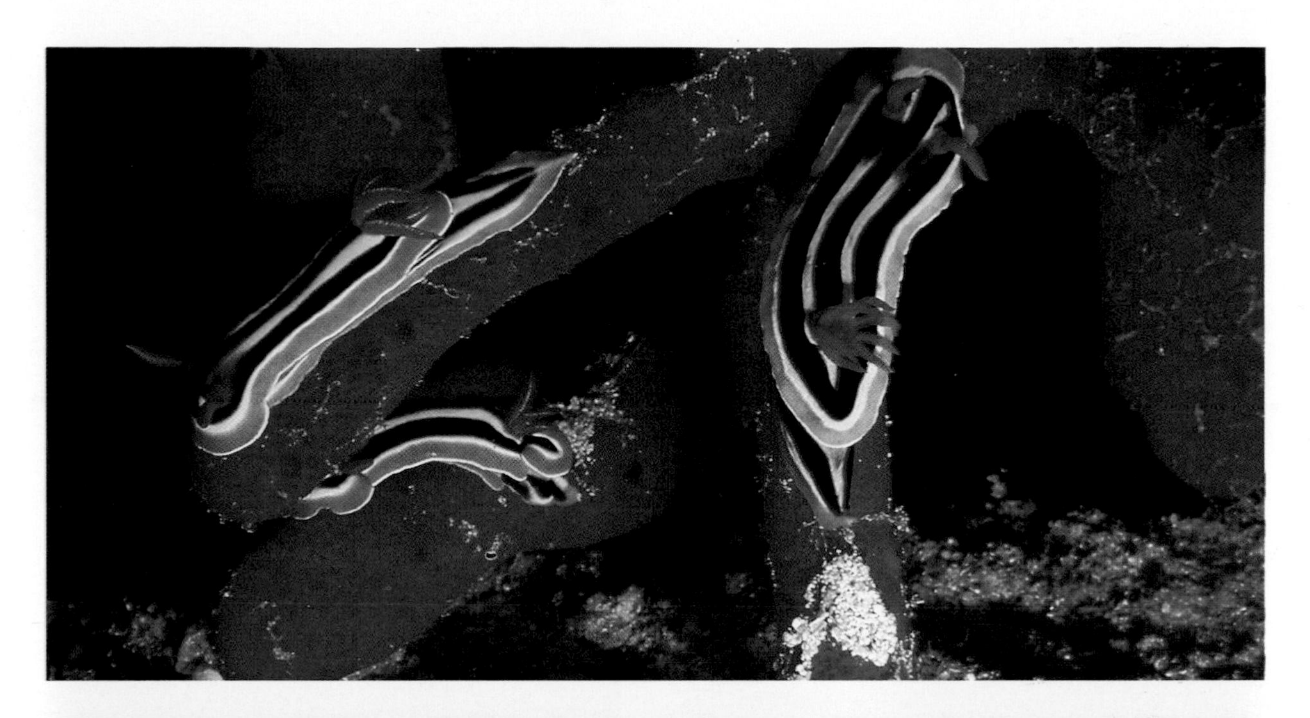

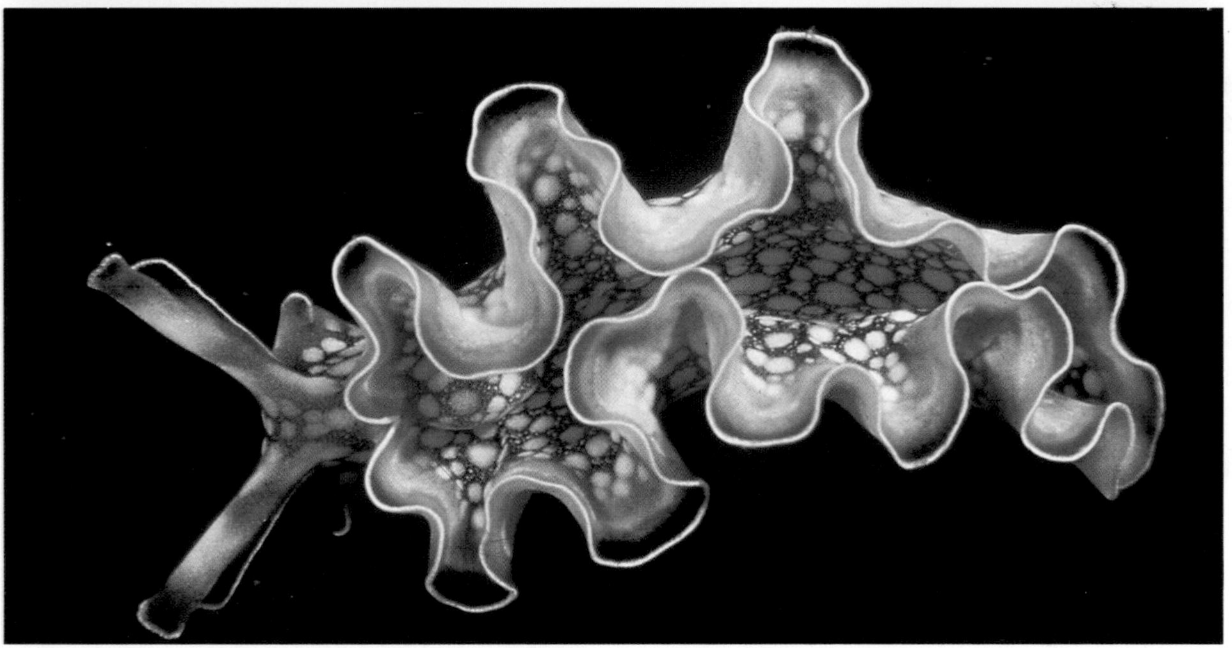

Pyjama Nudibranch (*Chromodoris quadricolor*)

TOP. Photographed on fingers of red sponge on the Great Barrier Reef, these nudibranchs are found in warm waters throughout the world. They have a row of stinging glands on the edges of their mantle and are distinguished by their curious habit of continuously twitching their gills.

Floating-Garden Slug (*Tridachia crispata*)

BOTTOM. This extraordinary creature lives in symbiosis with green plant cells which it bites from a piece of seaweed and stores in its transparent body. It then crawls near the surface, absorbing sunlight. The plant cells provide it with a constant supply of sugars as a result of photosynthesis. And so it has no need to eat at all.

SIX

PROPAGATING THE SPECIES

The reproduction of animals is a source of never-ending invention and complexity in nature. The ferocious but rarely fatal male rivalry fights that establish in many species who is powerful enough to mate, the elaborate courtship dances and rituals, the varied and often bizarre forms of copulation, the incubation and rearing of the young: each stage of the reproductive process is vital to the survival and multiplication of a species. Great Slugs spend up to an hour and a half in slimy foreplay before hanging in the air in a spiral embrace. Female deep-sea angler fish use the males of the species as portable sperm banks, while aphids dispense with males altogether for generations at a time, reproducing by virgin birth. Rousette bats form "maternity wards" consisting exclusively of pregnant females or those who have just given birth, while amongst the emus, it is the male who builds the nest, incubates the eggs and rears the young.

Waved Albatross (*Diomedea irrorata*)

At the start of the breeding season, the females of this species have to actively seek out the males and encourage them to begin courting. The ritual that follows includes beak duelling, beak clapping (shown here), neck waving, bowing and "sky pointing". The male finally subdues the female and seizes her neck in his beak during copulation. The Waved Albatross is the only species of the albatross family to live in tropical latitudes. This pair was photographed on Espaniola Island in the Galapagos.

Shingleback Lizard (*Trachydosaurus rugosus*)

ABOVE. These slow-moving lizards, shown here during courtship, live in southern Australia. Their name derives from their scales, and the stumpy tail is thought to act as a fat store. Perfectly adapted to desert life, they have short legs which they use to "swim" through loose sand and transparent discs in their eyelids so that they can see where they are going even when their eyes are closed to keep sand out. (See also page 129.)

Great Slug (*Limax maximus*)

OPPOSITE. After crawling up an overhanging branch or wall, these hermaphroditic creatures circle each other for 30–90 minutes while exchanging mutual caresses with their tentacles and secreting a great quantity of mucus which eventually forms into a twisted patch of glue-like consistency some two inches in diameter. The slugs then entwine themselves corkscrew fashion, detach themselves from their base and hang by a twisted cord of the thickened mucus 8–15 inches long. Still twisting in their embrace, they extrude from under their heads their blue, club-shaped penis sacs which then extend into a fan shape at the tips. The two "penial masses" are then intertwined in a tight spiral, the upper coils expanding to form a kind of umbrella. When this is achieved, sperm transfer takes place and the slugs untwist themselves, crawl back up the mucus cord (which one of them eats), slide down the tree or wall and return to their everyday life in such damp environments as the undersides of fallen logs.

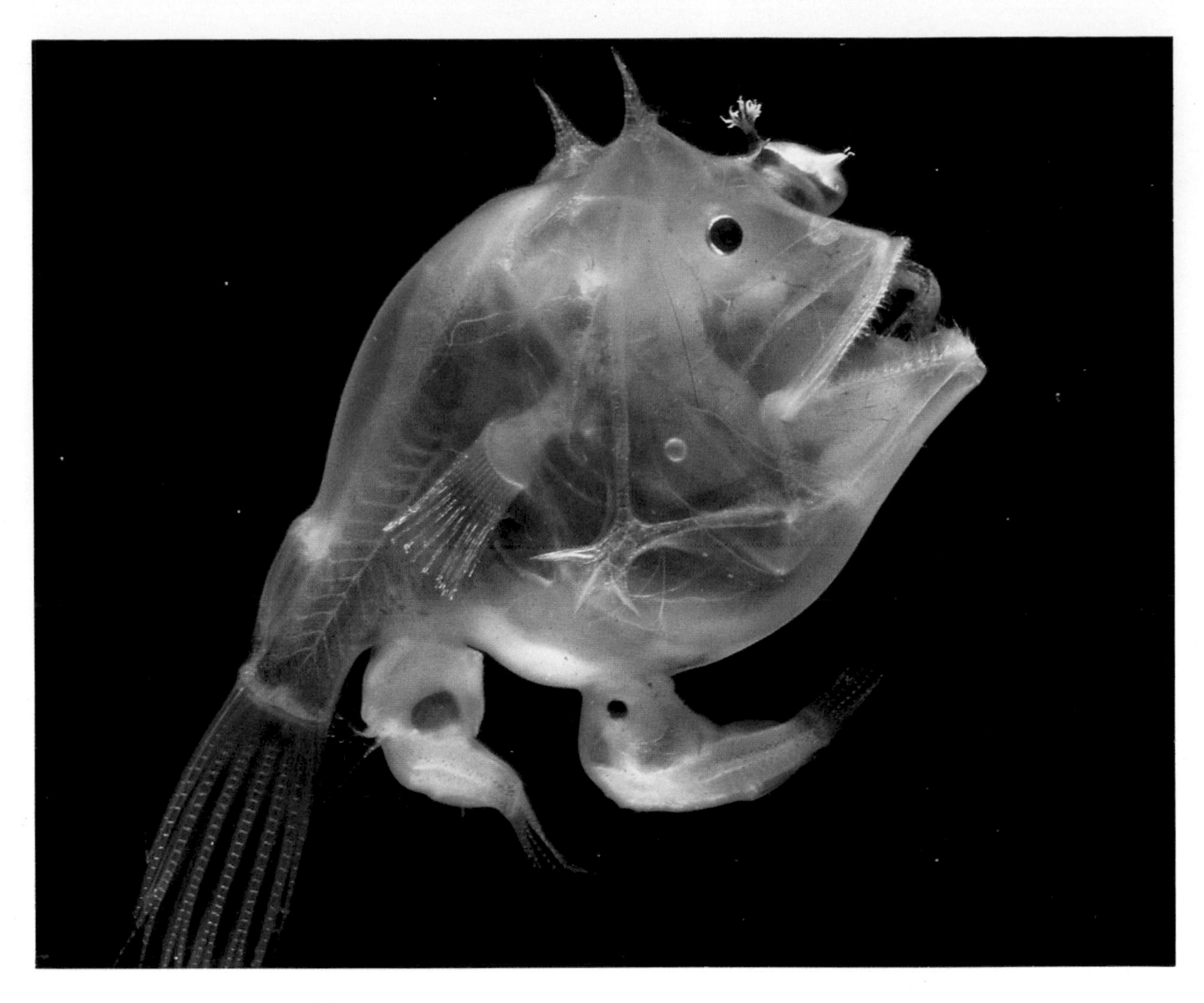

Deep-Sea Angler Fish (*Edriolychnus schmidti*)

ABOVE. To survive in the total blackness of the deep, many deep-sea anglers have evolved a reproductive strategy in which the sexes make contact immediately after birth and stay connected for life. Shortly after the eggs hatch, one or more males attach themselves to a female, sinking their jaws into her and never letting go again. Over time, the skin of the growing female spreads over the male and her blood system incorporates his. His mouth disappears completely and he becomes a mere sex appendage or sperm bank. The photograph shows two males less than three-quarters of an inch long attached to a female some three inches long. Her semi-transparent skin reveals much of her internal structure and organs.

Aphids (Aphidinae)

OPPOSITE. Humans wage a non-stop global war on the many species of aphids that attack crop plants, but their astronomical powers of population increase mean it is an endless battle. First generation female aphids lay their eggs in the autumn. In the spring nymphs hatch out, all of them females who lose their wings. As each becomes sexually mature, she becomes the ancestress of a whole new colony, producing live young (top) by virgin birth. This new generation also consists entirely of females, and so the process continues through several further generations until, at the end of the summer, the first male offspring appear. These mate with both the original females and their innumerable female descendants. The generation that results from this sexual mating are wingless females who lay the eggs which, next year, will start the whole cycle anew.

Larger Emu (*Dromaius novaehollandiae*)

Up to six feet tall, this is the second largest bird after the African ostrich. It can run at 30 m.p.h. and is an able swimmer. The female lays a number of wrinkled dark green eggs, weighing over a pound each, between December and April. But it is the male who builds the shallow nest of leaves, grass and bark, incubates in it up to 25 eggs from several hens, and raises and cares for the young, as in this picture. Three-toed and flightless, emus are found today only in the Australian bush, having been wiped out around all centres of population because, it is claimed, they drink water meant for sheep and cattle, trample fences and eat crops.

Desert Water-Holding Frog (*Cyclorana cultripes*)

Like other desert animals, this frog's life is dominated by its need to get and conserve water. When the ground begins to dry out after rain, it digs a burrow and begins to form a cocoon about itself. The process takes two to five weeks, during which the frog remains motionless, with eyes closed. The cocoon covers the entire body, including eyes and mouth (1), and reduces the animal's water loss to five times less than normal. When the next rain comes, the frog contracts its body (2) to loosen the cocoon and then pushes it forward with its hind legs (3) and front legs (4). It opens its mouth wide (5) and swallows the cocoon skin (6/7). With the cocoon gone (8), it is ready to return to the surface and enjoy the wet weather.

Cyclorana is found in northern Australia, is two inches long and has a mating call like a moaning growl.

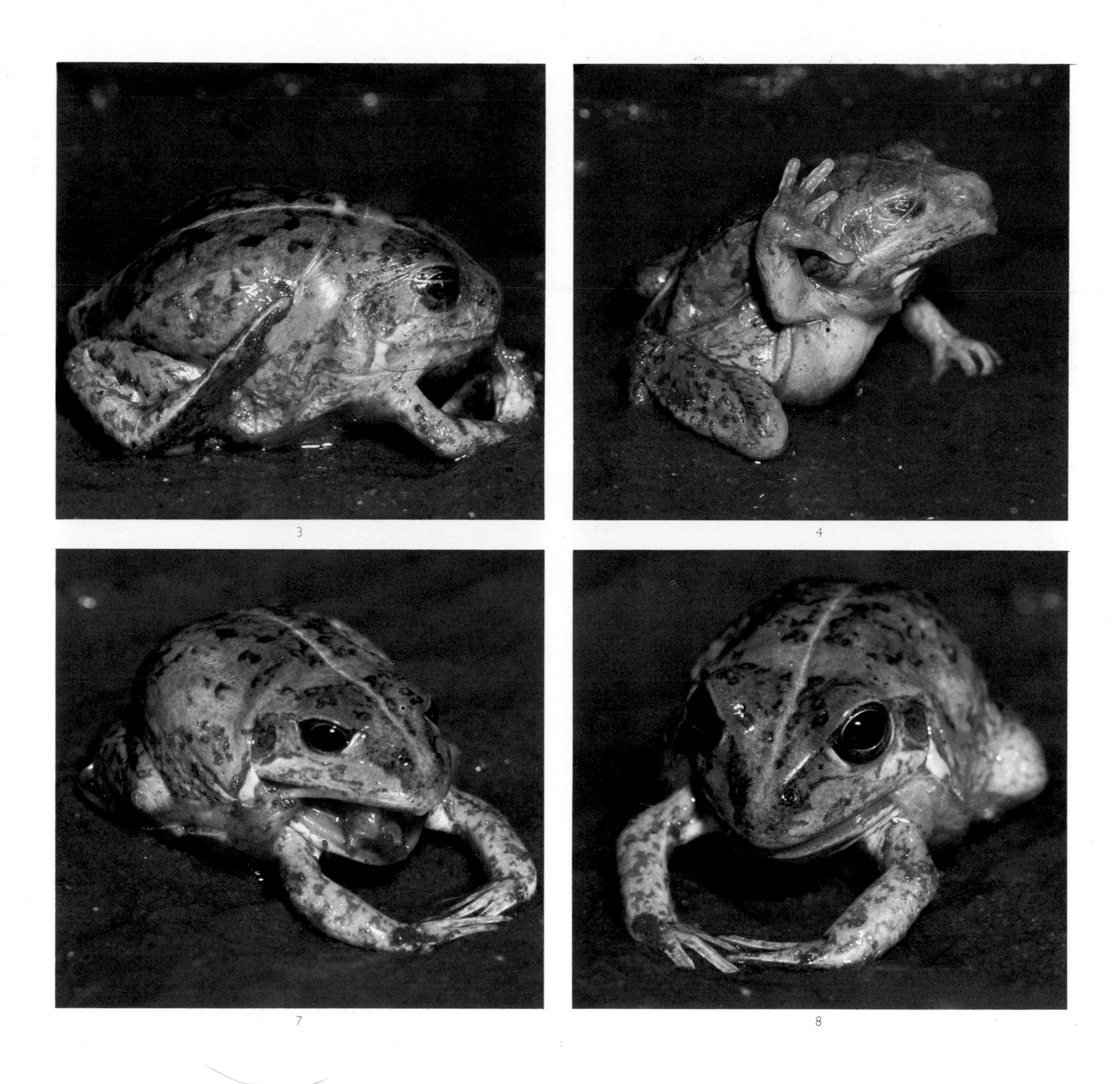

3 4 7 8

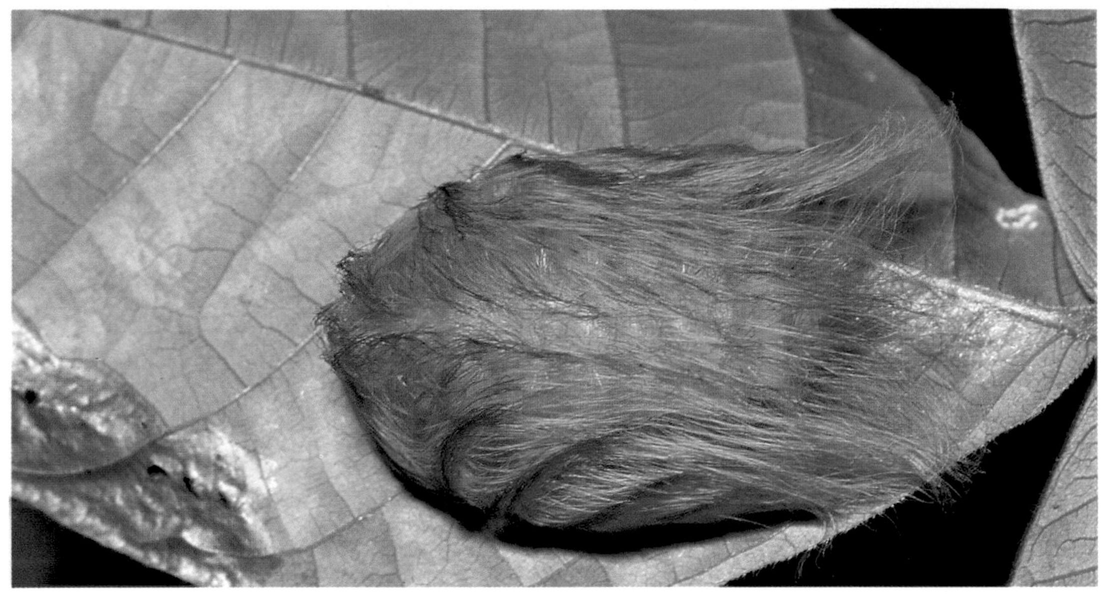

Tortoise Beetle (*Aspidomorpha tecta*)

TOP. The source of this little insect's striking similarity to its namesake is the iridescent gold pattern, like a tortoise's shell, on its back. The colour is in fact produced by the beetle's body fluids and disappears as soon as it dies. The larvae of the species have their own distinctive camouflage technique. They collect their own moulted skins and faeces and carry them on their backs, where they are held in place by long spines so as to form a kind of protective body casing.

American Flannel Moth (*Megalopygae crispata*)

BOTTOM. The fleshy caterpillar of this moth is just over an inch long and is found in the north-eastern United States. To protect themselves from predators, they have evolved a thick, hairy covering of poisonous spines which inflict a painful sting. They feed on oak, maple and sycamore and build themselves tough oval cocoons which are fastened securely to the side of a twig. They hibernate inside these until July, when they emerge through a flat, circular lid at one end of the cocoon.

Ringed Snake (*Natrix natrix*)

ABOVE. When threatened, this common European grass snake performs a brilliant imitation of death. It flicks out its tongue, hisses, releases the contents of its stink glands, which are said to smell like a mixture of garlic and mouse manure, defecates or regurgitates its food, dangles out its tongue and finally becomes completely rigid. If handled in this condition, there is no response at all. When not playing dead, the Ringed Snake usually lives near bodies of water, being an adept swimmer. Perhaps because it does not bite humans, it has since antiquity enjoyed the reputation of bringing good luck. For this reason, bowls of milk were put out in some parts of Europe in the hope of attracting it.

Praying Mantis (*Stenovates pantherina*)

OVERLEAF. This insect is named for its prayer-like posture when at rest. It will sit motionless for hours and then strike with lightning speed at its prey, seizing it in a deadly grasp with its saw-toothed forelegs. It is the only insect which can turn its head and look over its shoulder like a human. It drinks like a horse and washes its face like a cat. It is seen here in the threat display it gives when disturbed. There are many species of praying mantis and they are reputedly cannibalistic, the female devouring the male while in the act of mating. Some authorities dispute that this occurs except in unusual circumstances.

Giant Anteater (*Myrmecophaga tridactyla*)

OPPOSITE, TOP. From snout to tail, the Giant Anteater of Central and South America averages 9 feet in length. Its purpose-built body includes long tubular jaws ending in a small orifice from which a worm-like tongue can shoot out another two feet. Coated with sticky saliva, it is operated by a powerful muscle attached to the breast bone. The anteater has huge nails, which it protects by walking on its knuckles, and can open a termite mound with a single blow. It eats up to 30,000 ants and termites in a day.

Anaconda (*Eunectes murinus*)

OPPOSITE, BOTTOM. The longest reptile in the world, it feeds on mammals and birds that come to the river-bank to drink, killing its victims by constriction. Early Spanish settlers in South America called it "matatoro" – bull killer. It can certainly grow up to 30 feet, but there have been claimed sightings of specimens up to twice that length. Snakes are not thought of as affectionate creatures but during courtship Anacondas gently scratch their mates with their tiny vestigial legs.

Orang-Utan (*Pongo pygmaeus*)

ABOVE. Orang-Utan is Malay for "forest man" and these great apes are the only ones that spend all their lives in trees. As a result they have longer, stronger arms and fingers, and shorter and weaker legs. Constantly moving through the jungle in search of fruit, they live in groups of up to six individuals. Males are up to twice as large as females and weigh 165–220 lbs. An endangered species, there are less than 2,000 Orang-Utans left in the jungles of Borneo and Sumatra.

Narrow-Mouthed Toad (*Cophixsalis* sp.)

OPPOSITE. This tiny creature, less than an inch long, is an unidentified member of the diverse family of narrow-mouthed toads (Michrohylidae) which have spread from south-east Asia to Australia, northern China and the Americas. There are arboreal species, ground-dwelling species and some that burrow underground. Most have an egg-shaped body, a tapered head and no teeth. This one comes from Papua New Guinea and is seen inflating its vocal sacs.

Double-Crested Basilisk (*Basiliscus plumifrons*)

ABOVE. These bright green lizards are found throughout Central America and derive their name from the basilisk of fable, which was half cockerel, half snake and could kill at a glance. They live in bushes by riverbanks and, at the first sign of danger, drop to the water and scuttle over its surface at speeds of up to 7 m.p.h., using the scaly fringes on the long toes of their hind feet. This extraordinary characteristic has earned them two other popular names: the Jesus Christ Lizard, because they walk on water, and "Tetetereche", from the sound they make while doing this. Basilisks reach a length of some two and a half feet.

Black Swamp Weaver (*Amblyospiza albifrons*)

ABOVE. An inhabitant of African reeds, papyrus thickets and tall grasses, this is one of the 68 species of true weavers which are distinguished by their extraordinary skill in building nests on a genuine warp-and-woof system. The male typically does the major part of the nest-building, leaving the female to line the interior.

Koala (*Phascolarctos cinereus*)

OPPOSITE. The emblem animal of Australia, the Koala is not a bear at all but a nocturnal tree-dwelling marsupial, weighing some 35 lbs, which rarely descends to the ground. It feeds on the leaves of only 20 out of the 350 species of eucalyptus or gum trees and becomes so saturated with the oils that it smells of cough drops. Baby Koalas, while they are being weaned off their mother's milk and before they tackle the tough eucalyptus leaves, eat straight from their mother's anus – a habit known as coprophagy. In the 1920s, Koalas became a focus of the fur trade and were almost wiped out. Today they are a protected species.

NINE

BLUE FLIGHT

Underwater and in the air, body shapes are streamlined and adapted for speed, lift and power. Though divers and hanglider pilots may have an inkling of what it really is to swim and fly, the sea and the sky are three-dimensional media that are essentially alien to terrestrial creatures like ourselves. Life is dominated by movement. The environment is an ever-changing world of winds and waves, currents and temperature gradients. The animals on these pages glide, soar, streak and fly through realms beyond our experience.

Indian Flying Fox (*Pteropus giganteus*)

This animal is not a fox at all but a fruit bat. One of the largest of the 80 species of flying foxes, with a wingspan of four feet, it is mainly active at night. It crushes fruit for their juice, or eats them whole and spits out the kernels. It has also been observed eating blossoms and flowers. It uses its wings to fan itself when hot and as a wrap-around cloak when cold.

Magnificent Frigate Bird (*Fregata magnificens*)

OPPOSITE, TOP. The Man-Of-War Bird, as it is also known, is the fastest seabird, timed at over 95 m.p.h. It needs this speed since its principal food is flying fish. It also obtains food by chasing gannets until they regurgitate. With a wingspan of over seven feet, the Magnificent Frigate roams the southern oceans and, because of its infallible homing instinct, was used by Pacific islanders to carry messages.

Forster's Barracuda (*Sphyraena forsteri*)

OPPOSITE, BOTTOM. This shimmering school may look menacing but they are no danger to humans, growing only about a foot long. Their larger relatives can reach eight feet and have been known to attack swimmers. Forster's Barracudas are found in Indo-Pacific waters and feed off other schools of fish, migrating with them during the summer. Along with other barracuda species, they are poisonous to eat.

Bottlenose Dolphin (*Tursiops truncatus*)

ABOVE. Most of our present knowledge of dolphin behaviour comes from this species, the first to be kept in captivity. Pliny, the ancient Roman, wrote of them: "The dolphin is swifter than a bird and he hurls himself forward in the water faster than an arrow launched from a powerful machine." Ranging the Atlantic and Indian Oceans, Bottlenoses grow to 12 feet long, weigh 800 lbs, eat some 20 lbs of fish a day and reach speeds of up to 22 knots. Like other dolphins, their brains weigh more than ours, they care for each other and have few natural enemies except humans.

Californian Sea Lion (*Zalophus californianus*)

The popular view of sea lions is based on these agile and graceful animals which have become a staple attraction at many zoos and circuses. In fact, they differ from almost all other sea lions in being more slender, with a tapered dog-like head and no mane, and in being more at home on land. Up to seven feet long, over 50,000 of them live along the Californian coast, though individuals have been seen off British Columbia and even Japan.

TEN

EYES, SNOUTS & BEAKS

From the alien eye of the ancient Tuatara to the comically bulbous nose of the Proboscis Monkey and the startling blue tongue of the Shingleback Lizard, these pictures show some of the more extraordinary facial characteristics found in animals.

Tuatara (*Sphenodon punctatus*)

The large eye with vertically slit pupil indicates the twilight and nocturnal habits of this "living fossil", which is found only on 20 small islands off the coast of New Zealand. It is the sole survivor of an animal order that dates back 200 million years, before the era of the dinosaurs. Tuatara means "spine-bearer" in Maori, a reference to the crest along its back. It is olive-green and two feet long, does not breed until it is 20 years old and has an estimated lifespan of 100–300 years. It spends 90 percent of its time somnolent or asleep, hibernating for half the year. It performs best at a body temperature of 53.6°F (12°C), as compared to the 77–100°F preferred by other reptiles. So lethargic is the Tuatara that it has been observed to fall asleep even while eating. It also differs from all other reptiles, in that the male lacks a copulatory organ; transfer of sperm is achieved by males and females simply pressing their cloacal openings together. Tuataras eat insects, worms and snails and live in burrows originally built by petrels and shearwaters.

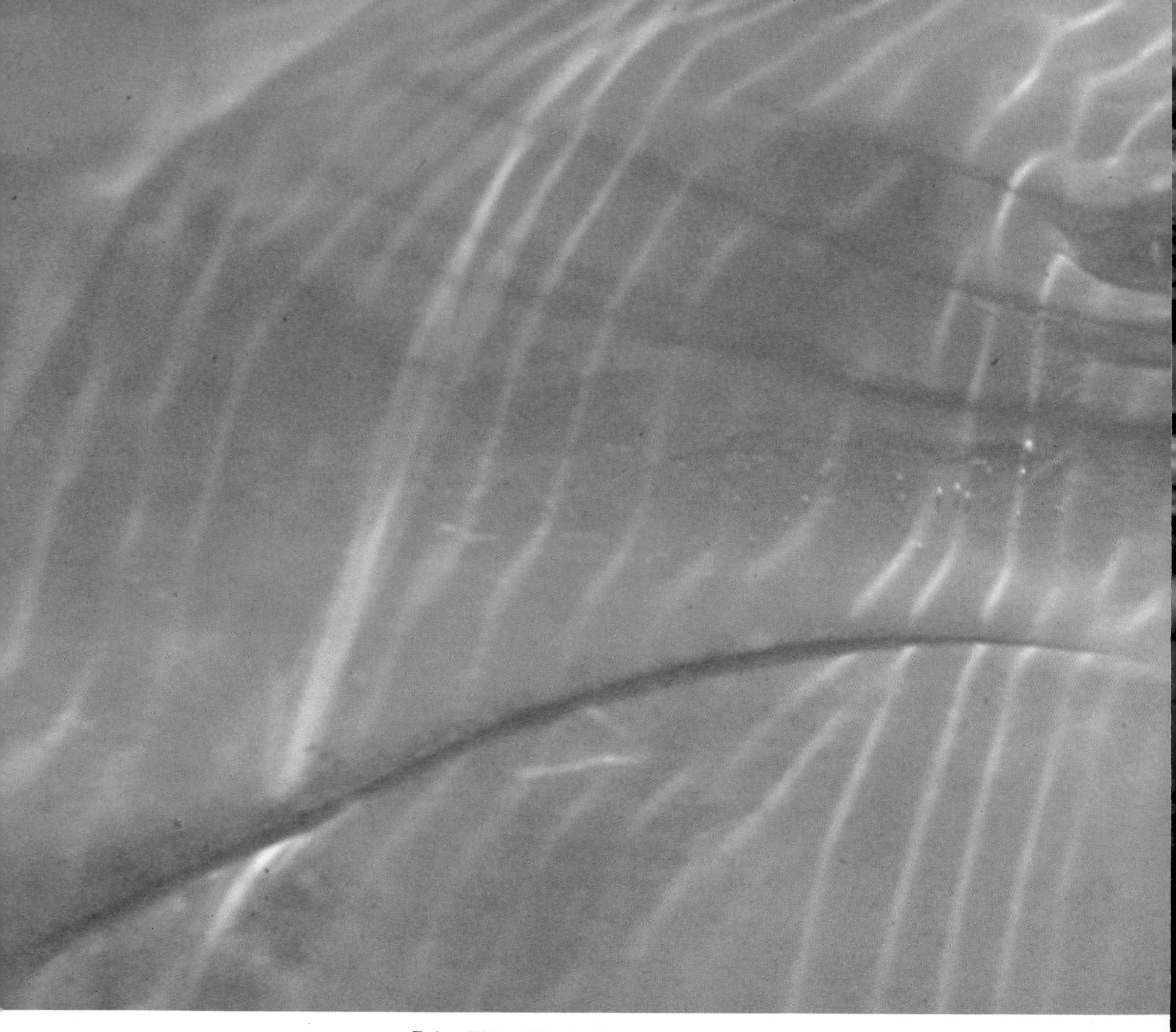

False Killer Whale (*Pseudorca crassidens*)

The eye of this whale is not much larger than a human eye at one and a half inches in diameter. It was photographed by Dr James Porter of the University of Georgia when 30 of these huge cetaceans, which grow up to 20 feet long, stranded themselves in 1966 on the Dry Tortuga Islands off the coast of Florida. Porter says: "Each whale involved in the stranding was identifiable at a glance by the highly individualistic facial characteristics and expressions." On this occasion only one of the whales died, the others being guided back into deep water with human assistance. An autopsy of the dead one showed that the cavities of its ears were packed with parasitical worms called nematodes, lending weight to the theory that whale and dolphin strandings occur when the animals' echo-location system is impaired. Another theory is that when one whale becomes stranded, perhaps for this reason, it sends out a distress signal which irresistibly attracts others to the scene, who then become stranded in their turn.

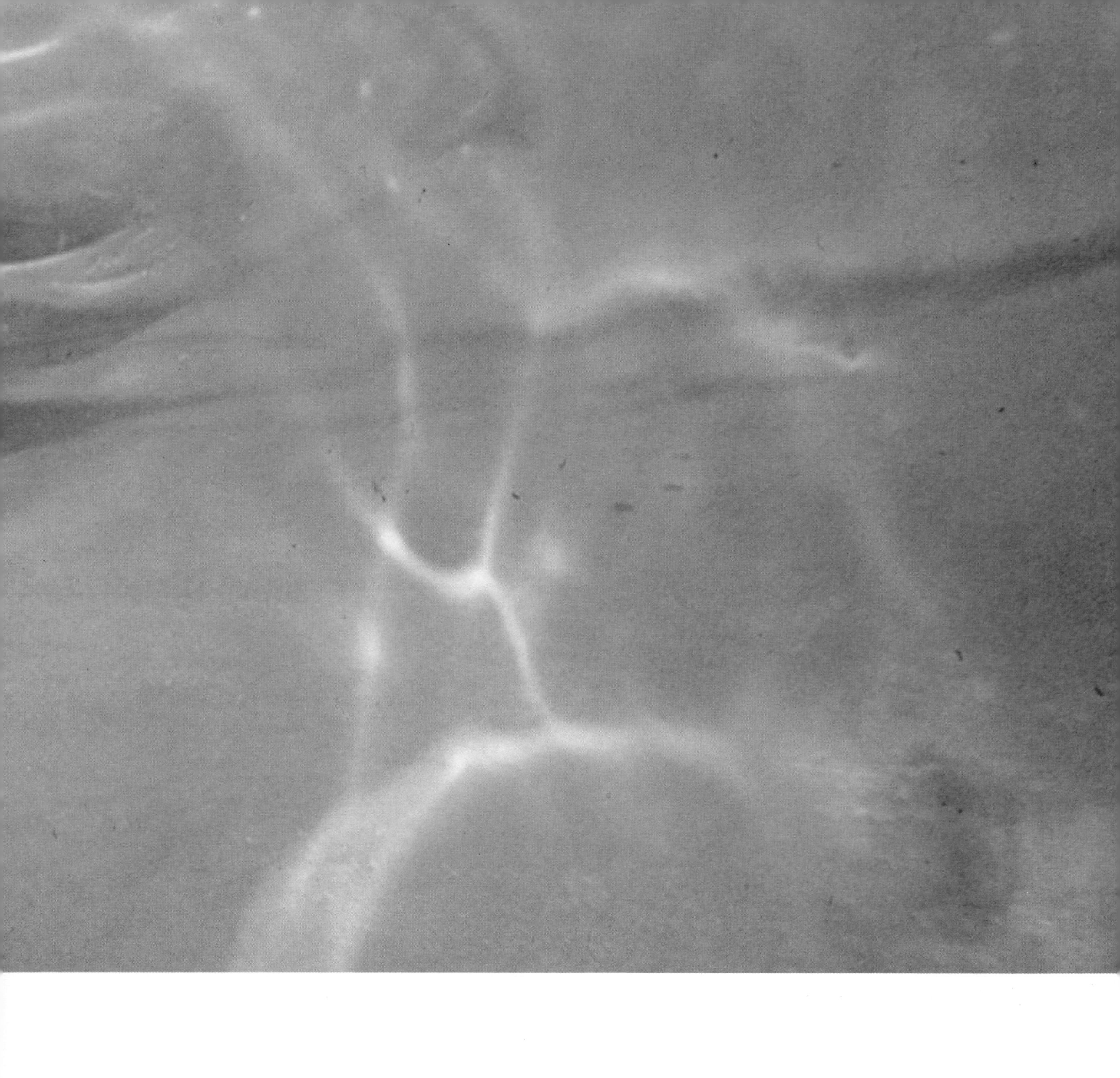

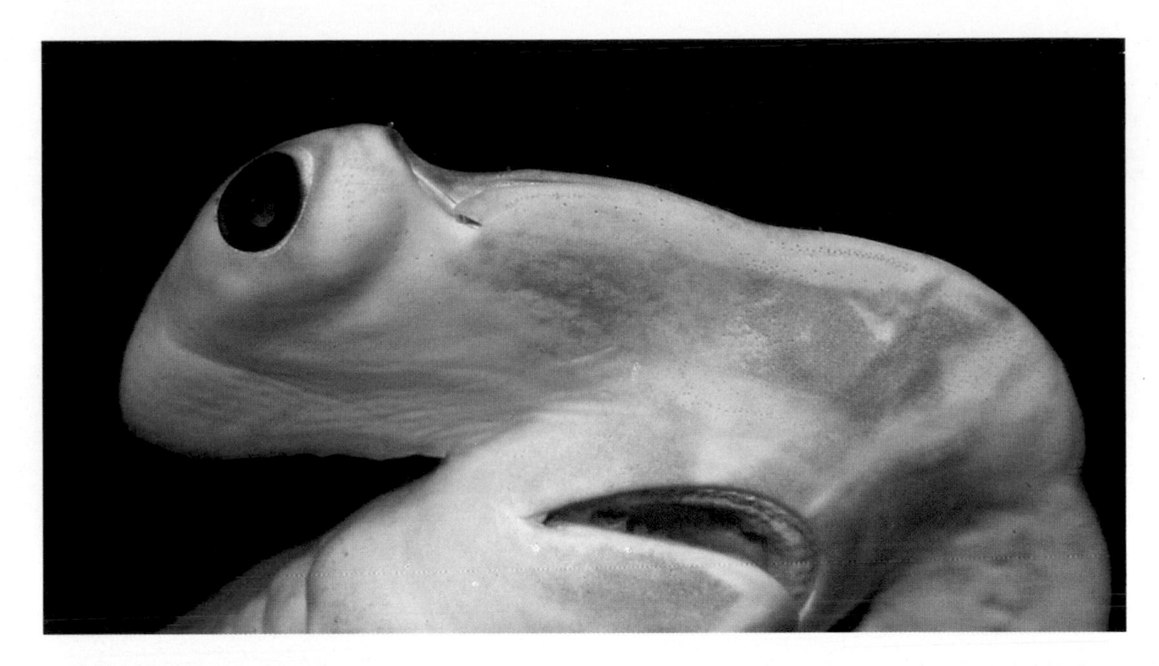

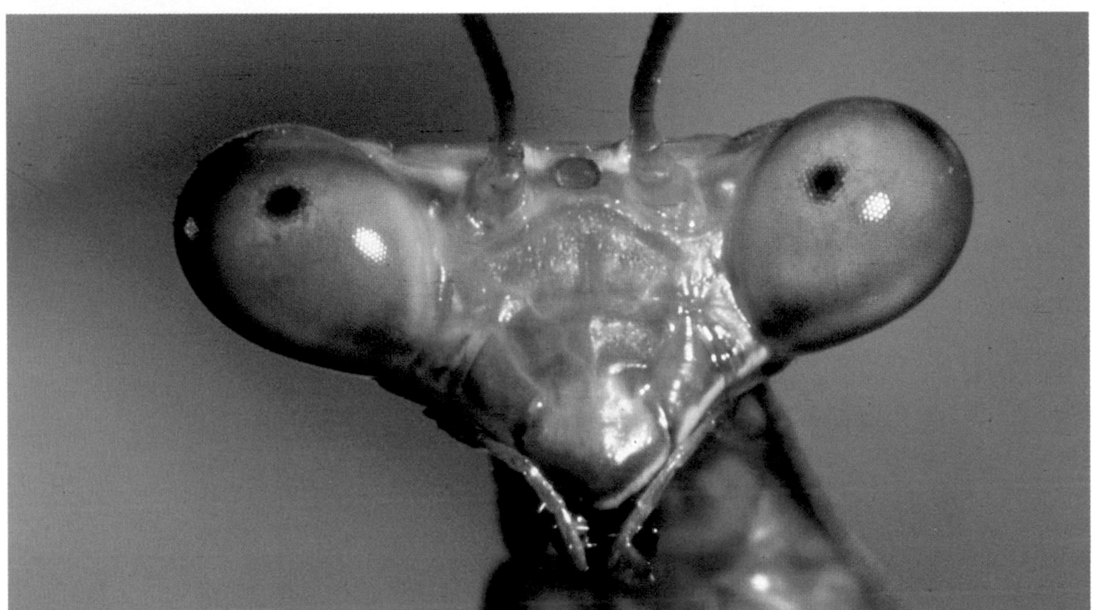

Common Hammerhead Shark (*Sphyrna zygaena*)

TOP. The unique head of this shark is extended sideways in flattened lobes which carry the eyes and nostrils at their tips. One theory suggests this shape provides the hammerhead with lift, another that it acts as the animal's rudder. A third opinion is that the shark gains an advantage from having its sensory pores spread over a wider area. The picture shows a juvenile hammerhead. Adults grow to 13 feet and are found in temperate and tropical waters. Dangerous and unpredictable creatures, they were fished commercially until the 1940s for the high vitamin A content of their liver.

Praying Mantis (*Pseudomantis albofimbriata*)

BOTTOM. Its head dominated by its huge mauve compound eyes, this is one of the 1,500 species of mantids found around the world. They are carnivorous insects related to cockroaches. Like cockroaches, the mantid females protect their eggs in special, papery egg capsules, each containing 100–200 eggs. (See also pages 94–95.)

Wagler's Palm Viper (*Trimeresurus wagleri*)

Visible between the nostril and the eye of this snake is one of the pair of pits which contain its heat-detecting organs. So sensitive are they that the animal can detect a change in temperature of as little as one three hundredth of a degree centigrade and sense the infrared radiation from any warm-blooded animal at a distance of up to two feet. In effect, it is equipped with thermovision as well as ordinary vision. This particular species of pit viper is a sluggish arboreal snake which preys on small mammals and birds and is found in Malaysia and the Sunda Islands.

Proboscis Monkey (*Nasalis larvatus*)

OPPOSITE. There is no mistaking the features of this peaceful, indolent inhabitant of the mangrove swamps and river banks of Borneo. The bulbous, three-inch long nose occurs only in the males and is thought to act partly as a sexual characteristic, partly as a voice amplifier for the animal's warning "honk". Proboscis Monkeys feed on mangrove shoots and the leaves of the padada tree. They spend much of the day sunning themselves in the trees in groups of 20 or more, but they are surprisingly agile when disturbed, making leaps of up to 25 feet. They are preyed on by Clouded Leopards and also by the local Dayak people, who eat their flesh.

Babirusa (*Babyrousa babyrussa*)

ABOVE. The two upper "horns" of this primitive, wild pig are in fact specialised canine teeth which pierce the upper lip and grow through the roof of its snout. They are too brittle to be of use as weapons and it is thought their value is in forming a protective screen in front of the upper parts of the face. Found only in forest swamps and reed thickets of the Celebes and Moluccas islands in south-east Asia, Babirusas do not root in the ground like other pigs, feeding on leaves, shoots and fruit. In older males the teeth curve round, occasionally growing back into the snout close to where they originate. Babirusas are about three feet long and two feet tall.

Southern Elephant Seal (*Mirounga leonina*)

OVERLEAF. The gaping mouth of this seal leads to the largest intestinal tract of any living mammal. Elephant seals are also unusual in having an extremely flexible spine which allows them to bend their back into a U or V shape. The mark of the huge adult males, who weigh over two tons, is their inflatable nose or proboscis. It grows gradually as the animal matures, becoming erect in the mating season. (See also page 81.)

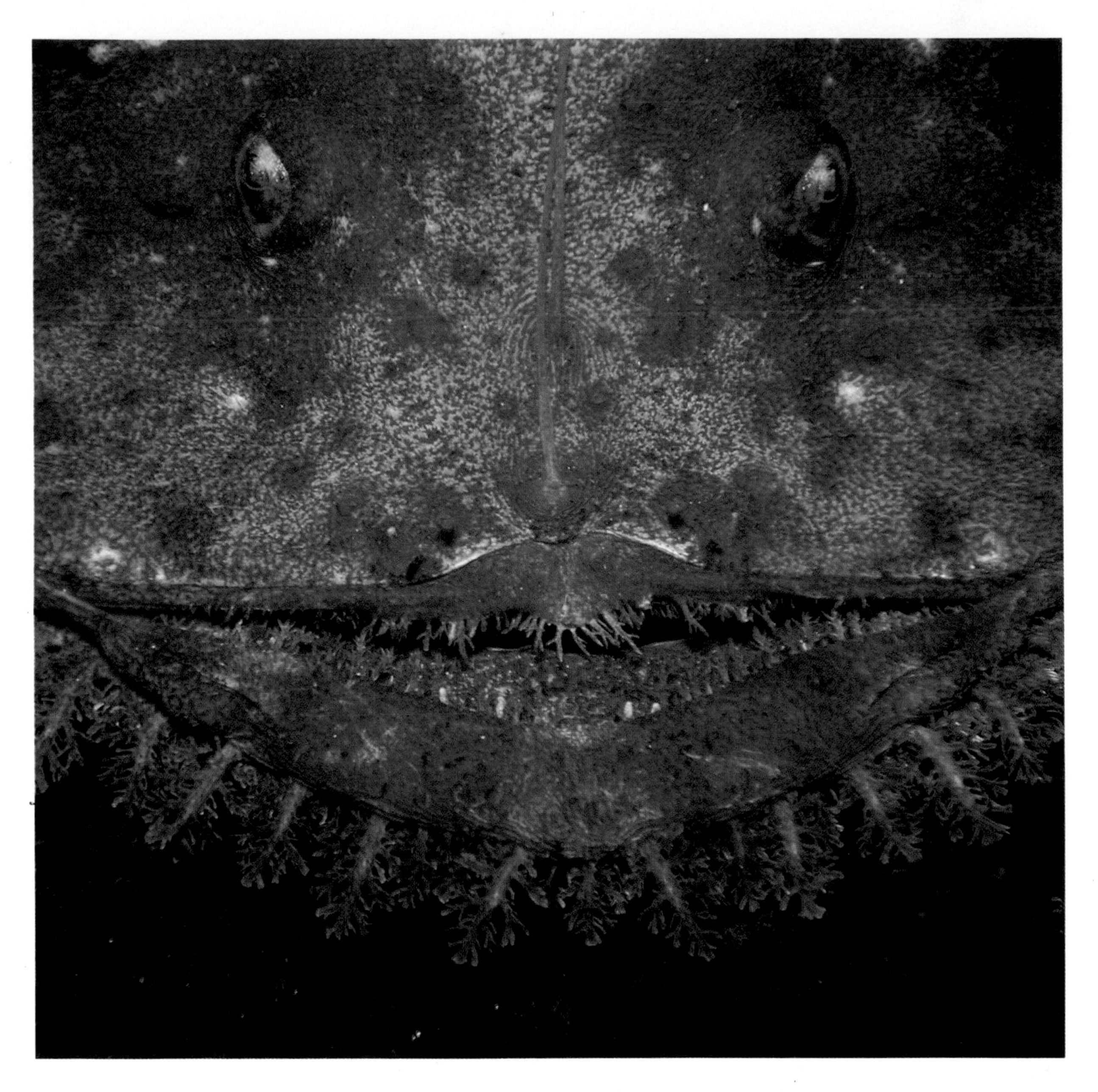

American Goosefish (*Lophius americanus*)

This strange, flattened, five-foot long fish has a mouth that runs the full width of its extraordinarily broad head and lips that are fringed with branching appendages. It spends most of its time lying on the seabed, blending into its surroundings. A line grows out from just above its upper lip, as can be seen in this picture, and at the end of it is a fleshy, worm-like lure which the American Goosefish dangles in front of its mouth to attract prey. Found all along the eastern coast of North America, these animals seize anything that investigates their bait. When they gather in shallow waters, they will also eat diving sea-birds.

Shingleback Lizard (*Trachydosaurus rigosus*)

TOP. The photograph shows the astonishing deep-blue tongue of this Australian desert reptile, which feeds on insects, carrion, flowers and fruit. It is described in more detail on page 74.

Cleaner Fish (*Labroides sp.*)

BOTTOM. One of the most extraordinary forms of mutual aid amongst animals is the role performed by the various species of cleaner fish, which keep the mouths and gills of other fish free from parasites. The Cleaner in this picture is seen at work on a Spotted Sweetlip. Marine biologists have discovered that there are special coral reefs, dubbed "cleaning stations", where fish of many different species wait their turn to be cleaned. Although many of the visiting fish normally prey on each other, the "cleaning stations" are apparently recognised as neutral ground and they wait their turn without disturbing each other.

Roseate Spoonbill (*Ajaja ajaja*)

OPPOSITE. Found in remote parts of Florida and Central and South America, it scoops material from muddy river bottoms and tastes it with the nerves inside its spoon-shaped bill. Living organisms are swallowed and inert matter washed out. Like scarlet ibises and flamingos, these birds lose the beautiful red and pink colouring of their plumage when they are kept in zoos, though it can be restored with food rich in carotene, the pigment found in carrots. Roseate Spoonbills are gregarious birds, but many of their colonies have been displaced by land development and they are also endangered by pesticides used in mosquito control.

Black-and-White Casqued Hornbill (*Bycanistes subcylindricus*)

ABOVE. The horny outgrowth or casque that surmounts this bird's enormous bill is in fact as light as sponge, being honeycombed with air chambers. The hornbill uses it as a defence against monkeys and snakes and as a resonating chamber to amplify its calls, which are like the braying of a donkey. An inhabitant of the rainforests of Africa's Ivory Coast, its omnivorous diet includes fruit, insects, eggs, nestlings and tree frogs. Unusually for a bird, it has well-developed eyelashes.

King Vulture (*Sarcoramphus papa*)

ABOVE. Third largest of the New World vultures after the Andean and Californian Condors, this huge voiceless bird, with its 6-foot wingspan, lives in the dense tropical rainforests of Central and South America. Because of its ability to instantly find the decomposing corpses of animals even when they are hidden in thick undergrowth, it is thought to be one of the few birds that locates its food by sense of smell. Little is known of its habits except that it does not build nests, laying its eggs in cavities between rocks or, quite simply, on the ground.

Shoebill (*Balaeniceps rex*)

OPPOSITE. Standing almost five feet tall and with an eight-foot wingspan, this relative of the storks and herons is one of the largest terrestrial birds. It lives in marshlands and along riverbanks in the Sudan, Uganda and the south-eastern Congo and has a reputation for nonchalant indolence. It spends hours at a time standing motionless and almost invisible amongst papyrus or other water reeds. Tiny aquatic creatures which it stirs up from the bottom of pools with its tough claws are its primary food, but it also eats fish, frogs and small reptiles such as softshell turtles. Usually silent, it occasionally makes a sound like shrill laughter.

ELEVEN

RIOT OF COLOUR

To human eyes colour is largely an aesthetic quality, but in nature it usually serves a specific purpose. Clown fish, for instance, are brightly coloured to attract predators who, in their turn, become prey when the Clown retreats amongst the stinging tentacles of a sea-anemone. The Flying Gurnard, by comparison, uses colour to frighten predators away, spreading wing-like fins covered with blue spots and stripes that make it appear much larger than it is. In other cases colour acts as a sex stimulant or a means of camouflage. It may even be an aberration, as in the case of the Mexican Leaf Frog pictured in this section, which is a rare blue mutation of a normally green animal.

Red-Blue-and-Green and Scarlet Macaws
(Ara chloroptera & Ara macao)

The macaws are the largest and most spectacular of the South American parrots. This picture shows a mixed group of Scarlet and Red-Blue-and-Green Macaws pecking at mineral-rich soil. The Scarlets, which grow three feet long, are distinguished by the bright yellow feathers on their backs. The imposing build and massive nutcracker bills of the macaws make them immune to predators. Their feet have two toes pointing forwards and two pointing backwards, which gives them a vice-like grip. The beak can also be used as a third "foot" and the birds sometimes hang from branches by it. Amongst the noisiest of jungle animals, macaws are kept tame in Indian villages throughout Central and South America. They are reputed to have saved a village on the Caribbean coast of Panama when they screeched furiously at the sight of a raiding party of Spanish conquistadors, arousing the Indians to action.

Atlas Moth (*Attacus atlas*)

A giant amongst moths, with a wingspan of up to 10 inches, the Atlas can be mistaken for a bird. It is found in south-east Asia and its wings have been described by renowned entomologist Walter Linsenmaier as "pictorial compositions of unique grandeur". Atlas caterpillars are covered with a powdery secretion and adorned with peg-like protrusions. Their cocoons can be used to make caterpillar silk, but the fibres have to be combed and spun like wool or cotton.

Flying Gurnard (*Dactylopterus volitans*)

Found in the warmer waters of the Atlantic and in the Red Sea, it is a bottom-living fish, a foot and a half long, and feeds on crustaceans. When alarmed, as in this picture, it spreads its huge pectoral fins in a startling display of colour designed to make predators think it is considerably larger than it really is. The Greeks described these fish as "swallows" but modern zoologists doubt that they can fly and ascribe reports of them doing so to mistaken identifications of flying fish.

Antarctic Isopod (*Antarcturus* sp.)

TOP. This delicate pink shrimp-like creature is an Antarctic representative of a group of extremely varied crustaceans that are found all over the world. The 4,000 species of isopod live in both salt and fresh water and have even conquered the land. There are isopods that parasitise fish and other crustaceans, transparent cave isopods, desert isopods, and white, eyeless ant isopods found underground in ants' nests. Amongst the better known are the many species of woodlice.

Cteniopus sulphureus

BOTTOM. These bright yellow beetles, about a third of an inch long, live primarily on flowers and are thought to feed on their pollen. Found in southern England and continental Europe, they are distinguished by a series of combs on the underside of their claws. The function of these is uncertain, but they may be an aid to movement. The larvae of *Cteniopus* burrow into soft, rotten wood where they feed on fungus.

Mexican Leaf Frog (*Pachymedusa dacnicolor*)

TOP. Mutations occur continuously in nature, as in the case of this frog whose distinctive blue is a radical departure from the normal green coloration of the species. *Pachymedusa* lives in the tropical lowlands of western Mexico where it is one of the few amphibians that remains active during the prolonged dry season. Its eggs, which are pale green, are deposited near water or in trees overhanging water. The tadpoles develop in shallow ponds.

Small Elephant Hawkmoth (*Deilephila porcellus*)

BOTTOM. An inhabitant of dry localities in Britain and Europe, it feeds on honeysuckle and spur-valerian and plays a major role in their pollination. Like other hawkmoths, it is active at dusk and after dark and is a powerful flyer. (See also page 151.)

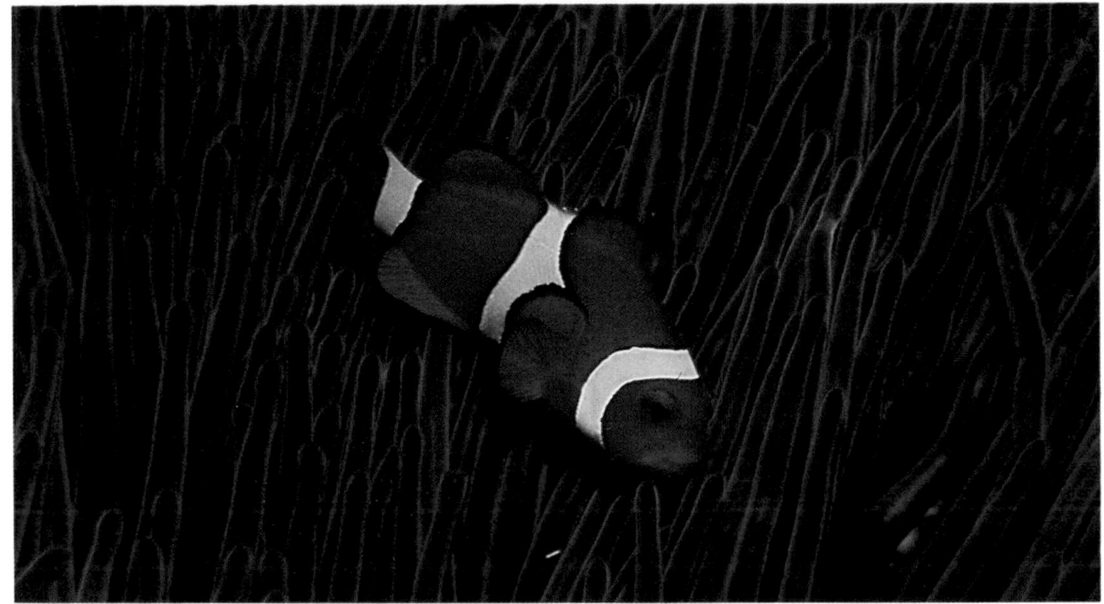

Clown Fish (*Amphiprion* sp.)

TOP & BOTTOM. There are 27 species of clown fish which live in symbiosis with 13 species of stinging sea anemones in the Indian and Pacific Oceans. The anemones benefit when a predator pursues a Clown too single-mindedly and gets stung as the latter retreats into the anemone's tentacles. The Clown uses the anemone as nest and fortress and shares in the food it helps to attract. Just how the Clowns avoid being stung themselves is a matter of conjecture. One theory is that they secrete a mucus which acts as a chemical cloak of invisibility. The top photograph shows *Amphiprion perideraion* with a large anemone. The bottom picture shows an unidentified *Amphiprion* species.

Malachite Kingfisher (*Corythornis cristatas*)

OPPOSITE. This magnificent bird is common throughout Africa south of the Sahara. It flies over rivers and swamps, searching for the fish, dragonflies and water insects it eats. When breeding, it favours rivers with banks in which it can excavate its nest holes.

Pachylis sp.

ABOVE. These multi-coloured creatures are not beetles but the largest known bugs, reaching a length of one and a quarter inches. Species of the *Pachylis* genus are found in Central and South America, but very little is known of their behaviour except that they feed on woody legumes like acacia. They are also thought to feature in the diet of some Amazonian Indians. This quintet was photographed in Peru.

Arrow Poison Frog (*Dendrobates auratus*)

OPPOSITE. Exquisitely coloured and patterned, its skin contains a poison which has a paralysing effect on the heart and nervous system. The Indians of Central and South America impale the frog on a stick and hold it over a fire until the poison oozes out. They then dip their arrow-heads in it and use them to hunt birds and monkeys. The tadpoles of these frogs are transported on their father's back between the tiny ponds of water that form in the leaves of bromeliad plants. Here they develop in a microenvironment complete with its own ready supply of insect food. The adults are less than an inch long.

Two-Lined Chameleon (*Chamaeleo bitaeniatus*)

This six-inch long reptile lives in the highlands of East Africa. It shows a marked preference for certain places where it likes to rest during the day or sleep at night. The females give birth to between 10 and 25 live young who measure just one and a half inches and take a whole year to become full-grown. Like all chameleons, the Two-Lined has a prehensile tail, eyes that can be rotated separately and a tongue like a coiled spring, which it can whip out to capture insects at a distance of some eight inches.

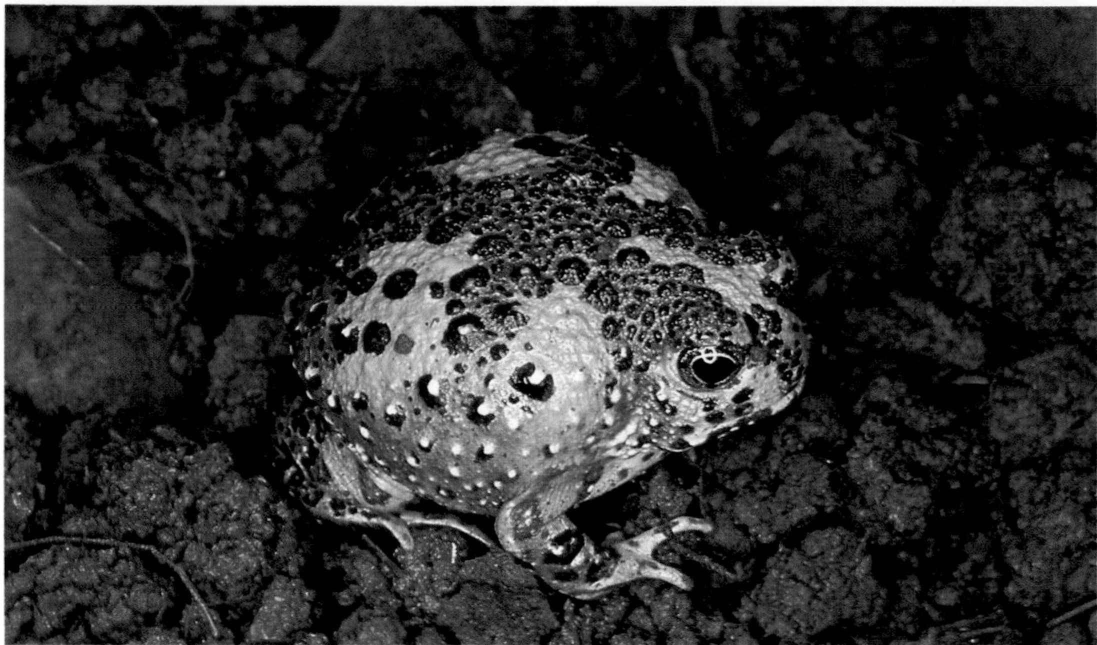

Papuan Tree Python (*Chondropython viridis*)

TOP. The yellow coloration of this snake shows that it is a juvenile. When it grows into a full-fledged, seven-foot long adult, it will be predominantly green. These pythons have adapted to life in jungle treetops. They have a prehensile tail which enables them to hang onto a branch as they shoot their entire body out to seize their prey. They live in Papua New Guinea and the Solomon and Aru Islands.

Holy Cross Frog (*Notaden bennetti*)

BOTTOM. Also known as the Catholic Frog, because of the distinctive black cross on its back, this is an Australian water-holding frog which is swollen after a shower of rain. The water is stored in its skin and it will now bury itself in the ground until the next rainstorm. (See also pages 90–91.)

Electric Ray (*Diplobatis ommata*)

ABOVE. This beautiful tiny ray, only six and a half inches long, is found off the Pacific coast of America from the Gulf of California to Panama. It has electric organs on either side of its head with which it can generate a shock of 50–60 volts. This facility is used not just to stun victims and deter predators but also as a sophisticated sonar device. The ray records minute changes in the low-voltage electric field with which it surrounds itself, and can thus detect extremely small obstacles and even differentiate between objects of the same size but varied electrical conductivity. It has, in effect, an electrical sense analogous to sight or hearing. Despite their fearsome equipment – some of the larger electric rays can emit a deadly 200-volt shock – these creatures are not aggressive and confine their predatory activities to crustaceans and other inhabitants of the seabed.

Yellow-Lipped Sea Krait (*Laticauda colubrina*)

OPPOSITE. Sea kraits are the least well adapted of the marine snakes and spend a considerable amount of time on land where they not only lay their eggs but can also be found sunning themselves. Like their cousins, the sea snakes, they have a flattened, rudder-shaped tail but their bodies are still cylindrical and have retained the muscles needed for sliding over the ground. This species reaches a length of three and a half feet and inhabits coral reefs of South Pacific islands. It is not aggressive unless irritated, but its poisonous fangs can give a fatal bite to humans. It lives chiefly on eels, which it detects by smell and overcomes with a series of bites all along their body. The venom blocks the eel's motor nerves and paralyses its muscles, though the heart may continue beating. The Krait then swallows the eel entirely, starting at the head.

THIRTEEN

DISAPPEARING ANIMALS

Merging into the environment is one of the most basic of all survival strategies, and the art of camouflage has been highly developed by animals of all kinds. Many species are not just the same colour as their background, but change colour as they move about and according to whether they are in shade or sunlight. Others, like the Sargassum Fish, grow appendages that resemble the vegetation amongst which they live. Mimicry of this kind has been perfected by the large number of insects who have evolved to appear almost indistinguishable from leaves, sticks and even stones.

Tawny Frogmouth (*Podargus strigoides*)

These nocturnal birds, related to nightjars, oil birds and whippoorwills, have enormous gaping bills, long wings and live on a diet of insects, reptiles and small mammals. At night they crowd together, but during the day they live in pairs, using their camouflage to blend into the trees where they sleep. If disturbed, they slowly stretch their neck and head, point their beak upwards, narrow their eyes to slits and freeze. This pair was photographed in Queensland, Australia.

Bush Cricket (Phaneropterinae)

ABOVE. The wings of this unidentified Costa Rican bush cricket are indistinguishable from the leaves amongst which it sits. About one and a half inches long, it is a member of the family of Katydids or Long-Horned Grasshoppers. As with other crickets, the males have at the base of their wing covers a special organ consisting of a stridulatory ridge, known as a scraper, and a file. When the file is pushed along the scraper, it produces the cricket's familiar chirping sound, which is used to attract females. Bush crickets are vegetarians and are distinguished by their very long antennae.

Sargassum Frogfish (*Histrio histrio*)

OPPOSITE. Some 800 miles due east of the Bahamas is the huge, egg-shaped area of ocean known as the Sargasso Sea, whose clumps of floating weed have been the source of much myth and legend. Amongst this weed lives the 11-inch long fish whose blotchy colouring and sharp spines perfectly mimic its plant environment. It even has white spots to match the casts of tube worms. Its pectoral and pelvic fins are modified to form prehensile "hands" which actually have "fingers" that clench the weed. When stalking the shrimp, pipe-fish, angler fish and worms that it eats, it swings hand over hand through the fronds of sargassum weed.

Leaf Mantis (Mantida)

OPPOSITE. Many of the 1,500 species of mantids have evolved with extraordinary modifications to their heads or bodies designed to help them blend into their environment. This unidentified specimen with the strange bell-shaped and leaf-like structure behind its head was photographed in Borneo.

Toad Grasshopper (*Trachypetrella andersonii*)

ABOVE. Found only in South Africa, it is hardly surprising that these short squat insects are known locally as "living stones". There is little information about them except that the females are wingless and twice the size of the males.

Spotted Scorpionfish (*Scorpaena plumieri*)

OVERLEAF. This remarkably camouflaged animal is one of a very diverse order of fish found in all the world's seas and oceans. Its large head is covered with bony armour, and spines at the base of its dorsal fins are attached to venom glands. The sting is very painful to humans and can have a lasting effect. Scorpionfish lie motionless waiting for a crustacean or another fish to pass nearby. They then pounce on it with one or more leaps, or open their mouths so rapidly that the prey is drawn into it by suction. This specimen was photographed off Abeyabu, one of the Palu Islands in the South Pacific.

FOURTEEN
CURIOUS CREATURES

There is a tradition of curious creatures, zoological oddities that have intrigued scientists or caught the imagination of the general public. One of the best examples is the Duck-Billed Platypus, an animal so strange that when the first complete skin reached the British Museum from Australia in 1798 the natural philosophers of the day were justifiably convinced it was a hoax. Who had ever heard of an animal with a duck's bill, reptilian shoulders and mammary glands? It was not until the first complete carcass arrived in England four years later that the zoologists accepted that the creature was genuine.

Duck-Billed Platypus (*Ornithorhynchus anatinus*)

OPPOSITE & OVERLEAF. The platypus belongs to the family of Monotremes, the most primitive of mammals, which have a single hole for both excretory and reproductive functions. A nocturnal creature, its 18-inch long body is covered with short, silky fur. The toes of its front feet are webbed to give more thrust in the water. The hind feet of the males are equipped with spurs attached to venom glands, making it one of the few poisonous mammals. The bill is not in fact as hard as a duck's and is extremely sensitive. The platypus uses it to scoop larvae, worms and molluscs from the bottom of the rivers and creeks in which it lives. It stores the food in cheek pouches until it surfaces, then crushes the shells between horny plates. It has a ravenous appetite and can eat more than half its own weight in a day. It usually stays underwater for about a minute, but can remain submerged five times as long if in danger. The female platypus digs an extensive burrow in the river bank, with a tunnel leading to a nesting chamber lined with gum leaves. Here she lays two or three eggs with soft, sticky shells about two weeks after mating. As shown overleaf, she suckles her young.

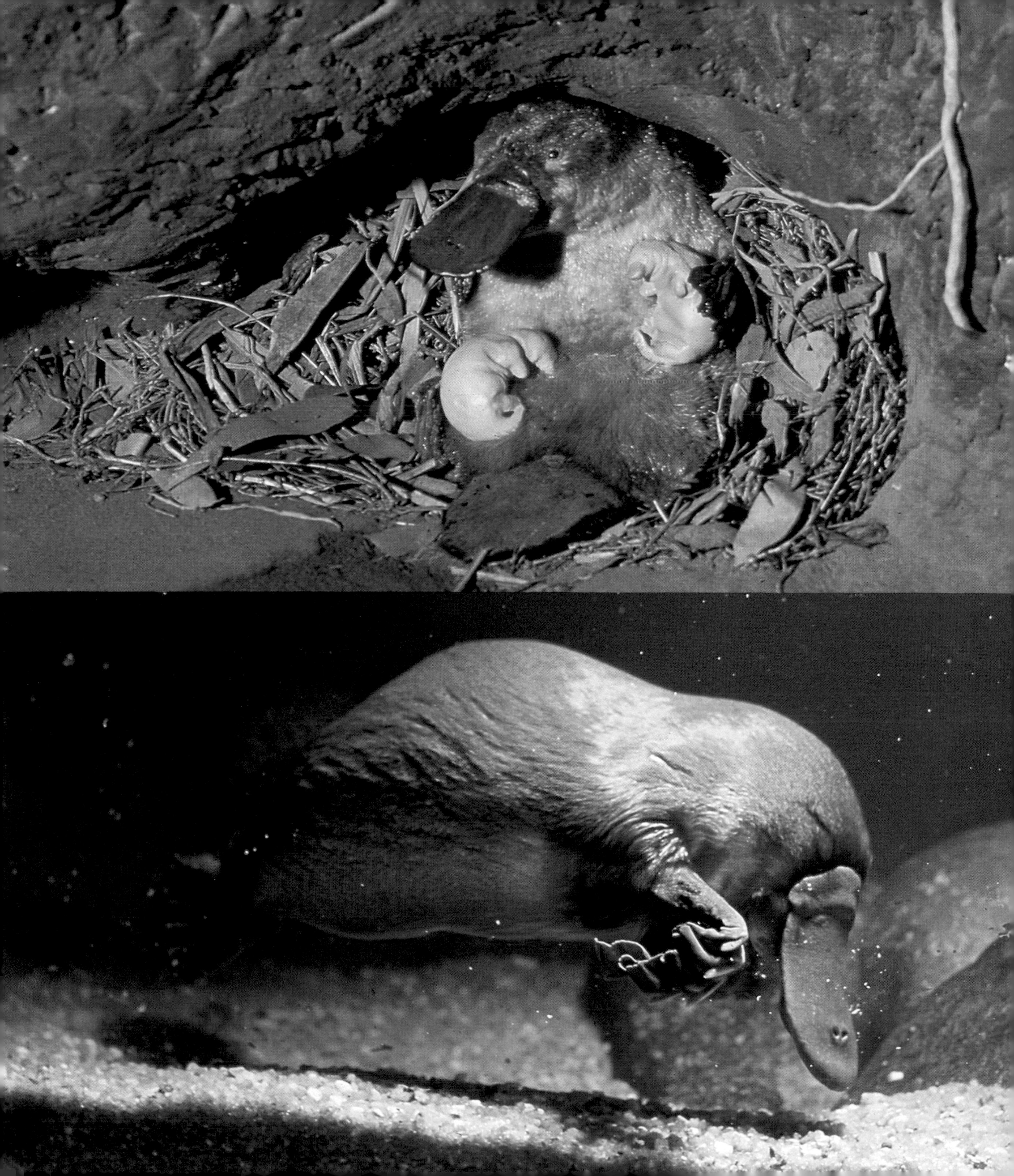

Coelacanth (*Latimeria chalumnae*)

ABOVE. The rediscovery of the Coelacanth caused a sensation. It was thought to have been extinct for 70 million years when a trawler caught one in 1938 and brought it in to the South African port of East London. It was named *Latimeria* after Miss Courtenay-Latimer, curator of the local museum, and *chalumnae* after the Chalumna River off which it was found. It was identified by Professor J.B.L. Smith, the authority on African fish, who said afterwards: "I would hardly have been more surprised if I met a dinosaur on the street." About 80 Coelacanths have been caught since. About five feet long and weighing 115 lbs, they have been described as "machines for reading the past backwards" because their heart and pituitary glands are at such an early stage of evolutionary development. They are extremely slimy to the touch because of the oil in their bright blue scales. Ironically, Coelacanths have long been known to the natives of the Comores Islands, off Madagascar, who eat them dried and salted and use their scales as a sandpaper substitute.

Aardvark (*Orycteropus afer*)

ABOVE. Five to six feet long, the Aardvark is a shy nocturnal animal with powerful claws which enable it to rip open termite mounds or dig itself out of sight in a few seconds. It lives in Africa south of the Sahara and is covered with bristly hair except on its head. It has a long sticky tongue for catching termites and is the only living representative of the order Tubulidentata, a classification resulting from its tube-shaped and rootless teeth which have no function and appear to be vestiges of an ancient form. Aardvark is Afrikaans for "earth pig".

Galapagos Tortoise (*Testudo elephantopus*)

OPPOSITE. There are several sub-species of these giant creatures, which grow up to five feet long and weigh some 500 lbs, on different islands of the Galapagos group. In each case the carapace is shaped slightly differently, one of the observations that led Darwin to his theory of natural selection. The vegetarian tortoises usually live in the warm, dry lowlands of their islands, but this creates a water problem which they solve by trekking regularly into the volcanic highlands to drink and wallow in muddy pools. They used to exist in massive numbers but have been reduced – on some islands to the brink of extinction – by competition with goats introduced by settlers and by slaughter for their meat. One source puts the number slaughtered since the discovery of the Galapagos at 10 million.

Hoatzin (*Opisthocomus hoatzin*)

One of the most primitive and unusual birds, it lives in the rainforests of Central and South America. Its name is Aztec in origin and is based on its cry. Hoatzins are reminiscent of Archaeopteryx, the reptile-bird of the dinosaur era, and although they can fly, they find it difficult and exhausting. The young have well-developed talons on their primary wing feathers which they use to crawl through trees or along the ground. The size of crows, Hoatzins eat leaves which they roll into balls and swallow whole. These leaf-balls are digested in their crop which takes up an extraordinary 13 percent of their weight. The contents of the crop are said to smell terrible and have given the Hoatzin its less complimentary name of "stinkbird".

Marine Iguana (*Amblyrhynchus cristatus*)

ABOVE & OVERLEAF. These iguanas from the Galapagos Islands are the only lizards which are at home in the sea, where they feed on algae and seaweed, excreting excess salt through special glands in their nasal cavities. Up to five and a half feet long, they emerge at first light from crevices in the black lava rock and divide their day between sunbathing to acquire body heat and lying in the surf, swimming and grazing when they become overheated. The picture overleaf illustrates the comment of naturalist Irenaus Eibl-Eibesfeldt when she first saw them: "One jump – and I felt as though I had travelled through centuries, back to an age in which dragons ruled the earth."

Himantolophus groenlandicus

ABOVE. In spite of its superficial similarity to *Linophryne* on the preceding pages, this angler is altogether larger at a length of two feet and has no barbel trailing from its chin. It is one of only three deep-sea anglers found in the colder waters of the North Atlantic and has occasionally been caught by trawlers working off Iceland and the Faroes. It usually lives at depths of 1,000–10,000 feet.

Deep-Sea Scaly Dragonfish (*Stomiata eustomias*)

OPPOSITE. A fearsome eel-like creature, it lures its prey with a long, branching bait in which pockets of symbiotic bacteria glow with a bright yellow light. Distributed in open waters, only some 10 species of these dragonfishes are known.

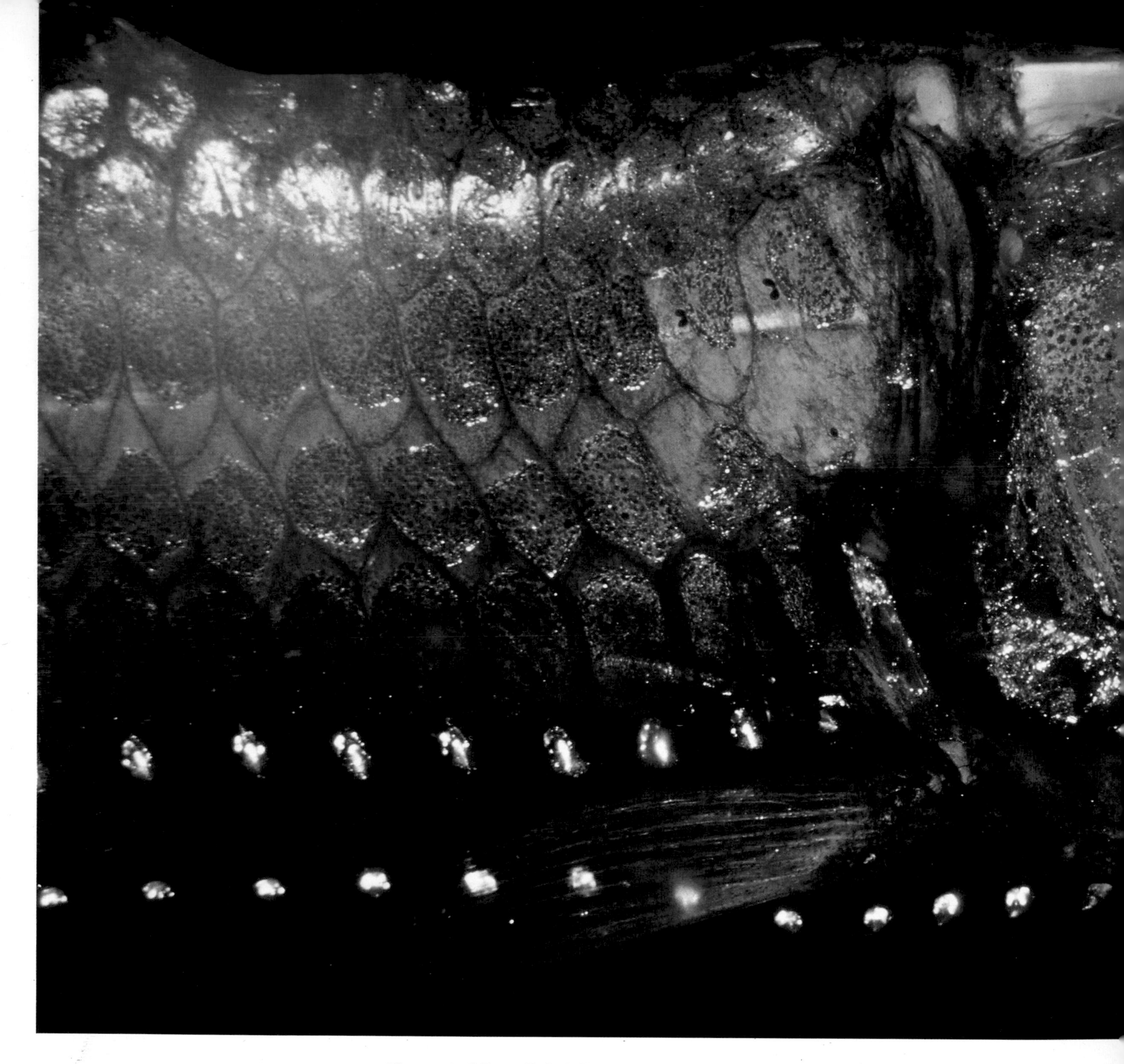

Sloane's Viperfish (*Chaulodius sloani*)

ABOVE & OVERLEAF. A slender, 10-inch long relative of the dragonfish, this creature is studded with bioluminescent organs and covered with deposits of opalescent crystals arranged in hexagonal patterns. It storms its prey with open mouth, throwing its lower jaw forward and rotating its upper jaw so that the needle-sharp teeth stick straight out from the head like spears to impale its victim. Movable throat teeth help to force down its gullet prey considerably larger than itself. An inhabitant of northern and southern oceans outside the tropics, it migrates vertically twice a day, spending the night near the surface and retreating to depths of 1,500–9,000 feet when the sun is up. It is named for its snake-like appearance and after Sir Hans Sloane, whose zoological collection forms the basis of the Natural History Museum in London and to whom the first specimen, caught off Gibraltar, was given for study. Its luminous display is shown overleaf.

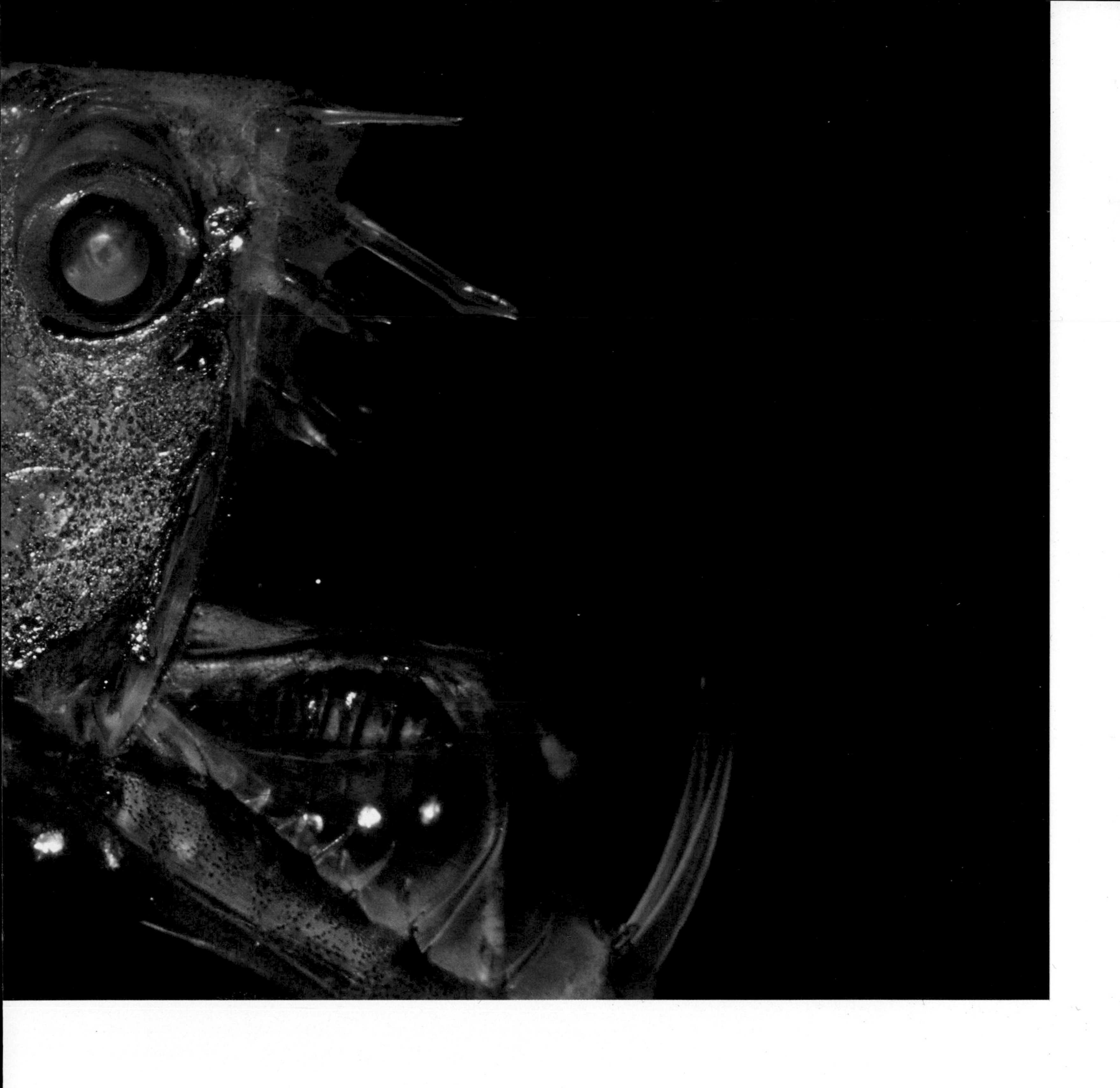

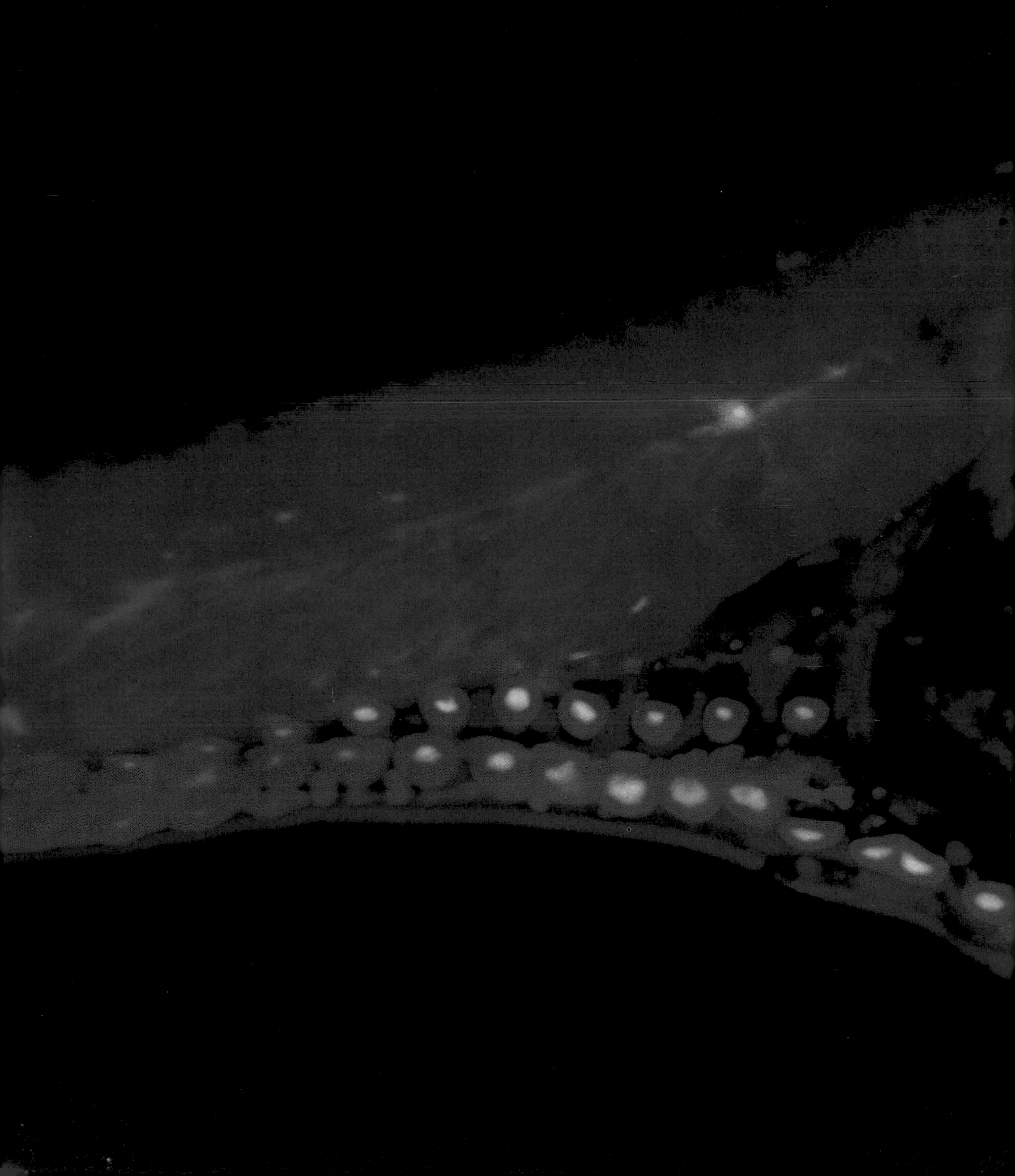

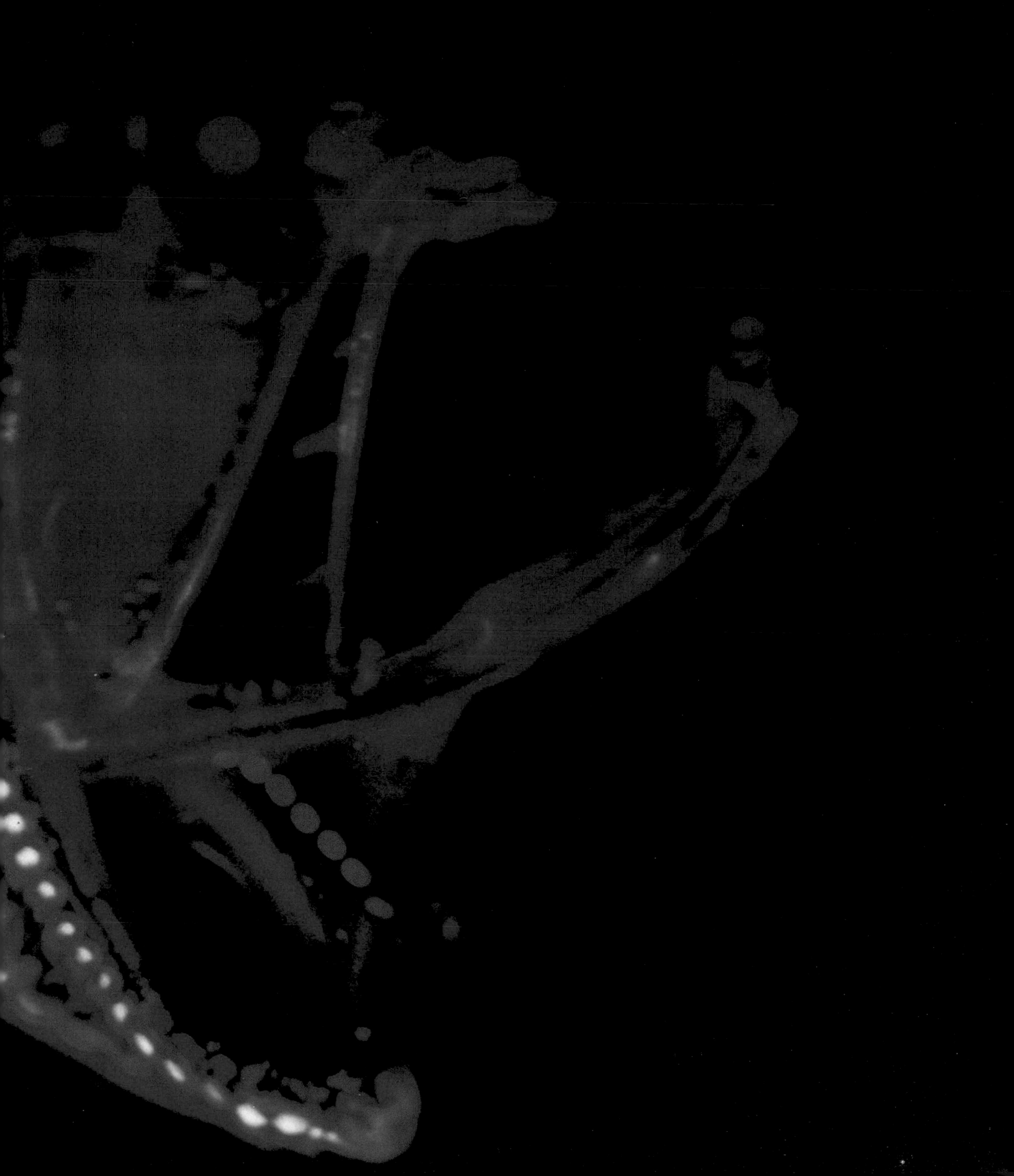

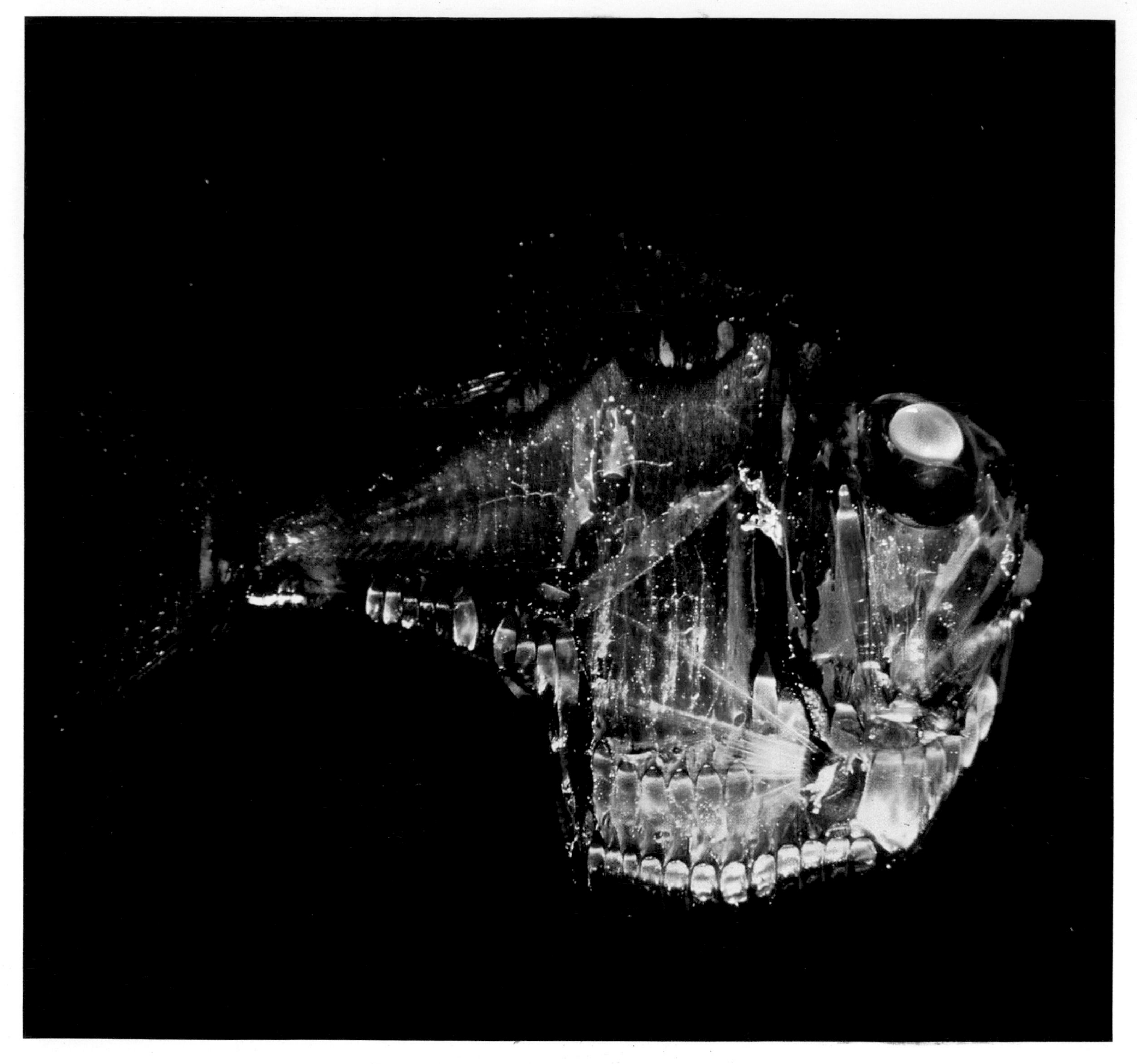

Deep-Sea Hatchet Fish (Sternoptychidae)

ABOVE & OPPOSITE. Some four inches long, these animals are found at great depth in tropical and temperate seas and are an important food source for tuna. They are named for their appearance, with the silvery body tapering to a sharp edge below the belly and coming complete with a thin tail like a hatchet's handle. *Argyropelecus olfersi* (above) has telescope eyes which are directed upwards. The three specimens opposite, swimming in formation, belong to an unidentified species. Hatchet fish have large luminescent organs on the underside of their bodies and glow brightly even at depth.

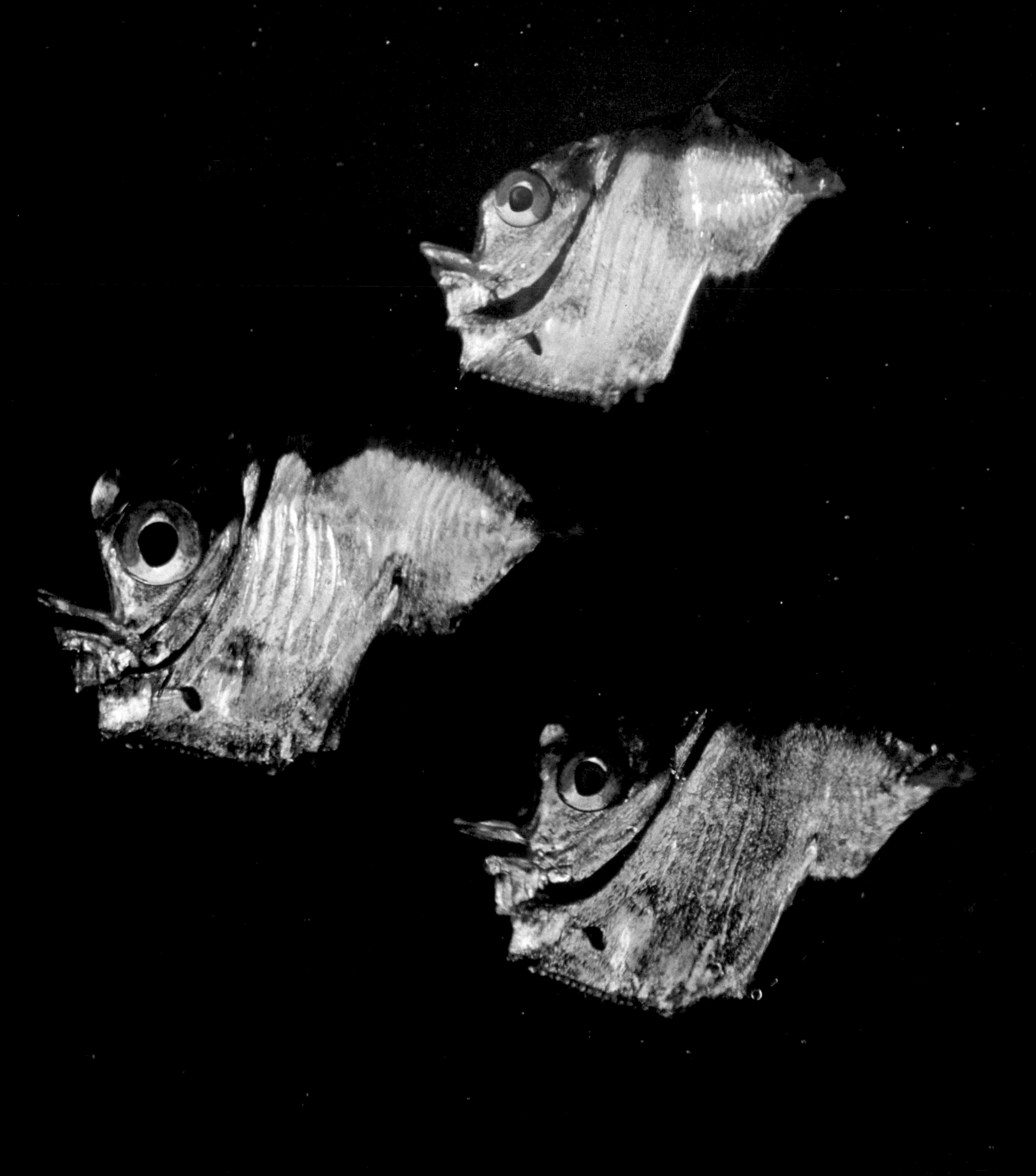

SIXTEEN

ANIMALS ALONE

These full-length portraits of individual animals are grouped together because we felt they expressed something about the extraordinary variety of nature – from the largest toad to one of the smallest crabs, from the Eastern Spiny Softshell Turtle to the Ganges River Dolphin. Variety is not just the spice of life, it is an essential requirement of evolution. When variety is diminished, within a species or within a family of species, so are the chances of survival. Some of these animals, such as the Black-Tail Prairie Dog and the Adelie Penguin, are in fact social animals, usually found with large numbers of their fellows. Others, including the Atlantic Octopus and the Pygmy Hippopotamus, are genuinely solitary creatures.

Atlantic Octopus (*Opisthoteuthis agassizi*)

Found in the North Atlantic from Ireland to the Gulf of Mexico, this octopus has a body about one foot long and eight arms that may extend several feet. In common with the other 200 or so species of octopus, it is jet-propelled, squirting water through an adjustable siphon, and confuses predators by clouding the water with ink. Octopuses can change colour faster than a chameleon and have 300 powerful cup-shaped suckers which they can clamp onto any solid surface. They can hang motionless in the water, walk on their arms and squeeze through any crevice they can get their hard, parrot-like beak through. The beak is used for prising open the shellfish they eat. Octopuses are solitary and intelligent creatures, far from the image of popular fiction.

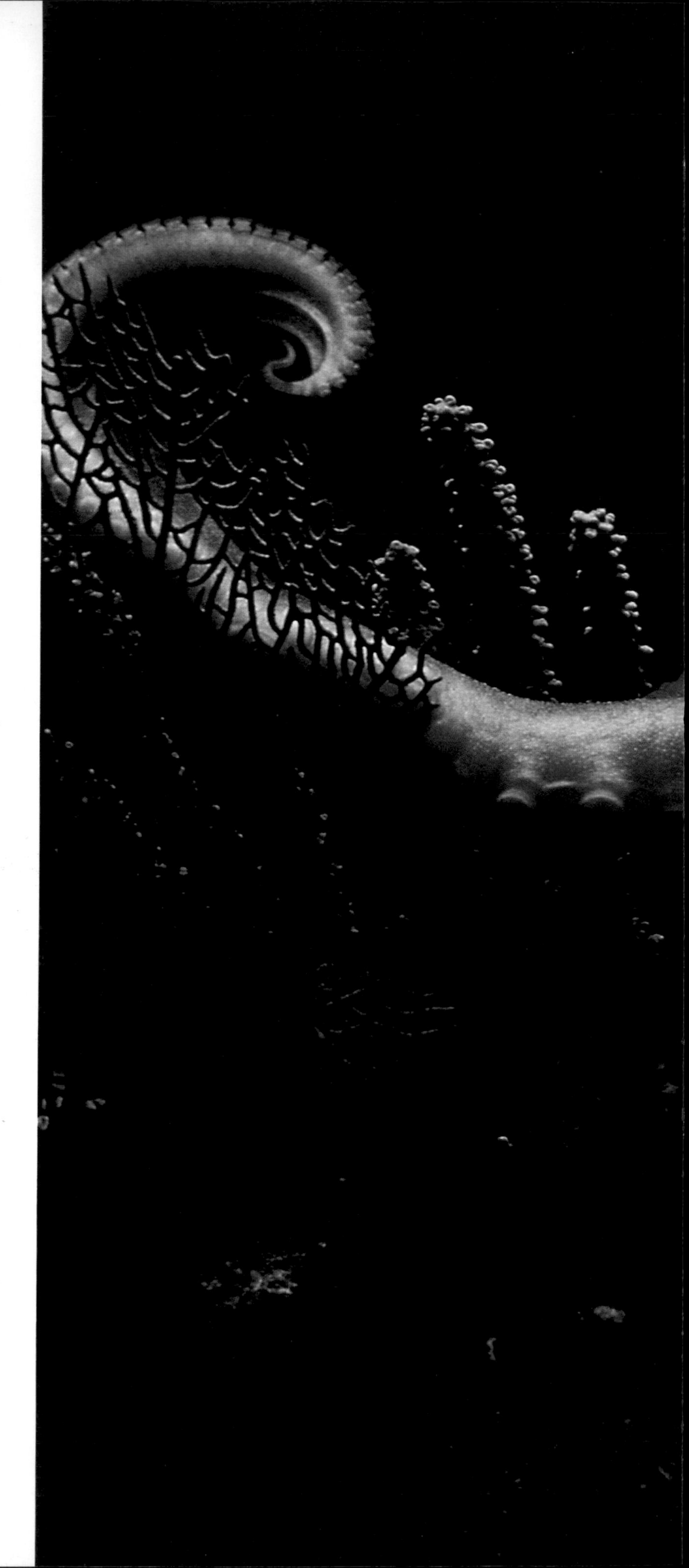

Black-Tail Prairie Dog (*Cynomys ludovicianus*)

ABOVE. Fast asleep on its haunches, this inhabitant of the American prairies is named after the black tip of its tail and its cry, which is like the bark of a dog. Prairie dogs form social groups called "coteries" and a number of these live together in "towns" – vast underground burrows which can cover an area of up to 160 acres.

Pygmy Hippopotamus (*Choeropsis liberiensis*)

OPPOSITE, TOP. This diminutive version of its better-known relative is less than five feet long and was first discovered in West Africa in 1849. Its hairless chocolate-brown body has a pink flush on its underside and is kept moist by mucus secretions. Rare and shy, the Pygmy Hippo feeds on fruit, plants and young shoots and sleeps in hollows in the river bank. Little is known of its habits and its endangered status means we may never find out.

Malayan Tapir (*Tapirus indicus*)

OPPOSITE, BOTTOM. This shy nocturnal inhabitant of the dense forests of Sumatra is an isolated relative of the other tapir species which are found exclusively in the Americas. Tapirs are related to rhinos and horses but are a more primitive form. The nose is extended into a sensitive trunk which serves as both a tactile and olfactory organ. The Malayan species is an excellent swimmer and stays close to water, feeding on leaves and aquatic plants. Its black-and-white marking helps to camouflage it by breaking up its body shape in the rainforest.

Marine Toad (*Bufo marinus*)

With a body length of almost eight inches and weighing in at just under three pounds, this is the world's largest toad. It is found throughout South and Central America and as far north as Texas. Also known as the Giant Toad, the Neotropical Toad and the Aga Toad, it can squirt poison at its enemies and its eggs are also poisonous.

Adelie Penguin (*Pygeoscelis adeliae*)

When spring reaches the Antarctic in September and October, the Adelie Penguins return to their "rookeries" to breed, often travelling more than 200 miles to reach them, navigating by sky signs. According to an elaborate, unchanging courtship ritual, the male offers a stone to a female and if she accepts it with a bow, he knows he has found his mate. Adelie Penguins are hunted by the predatory leopard seal.

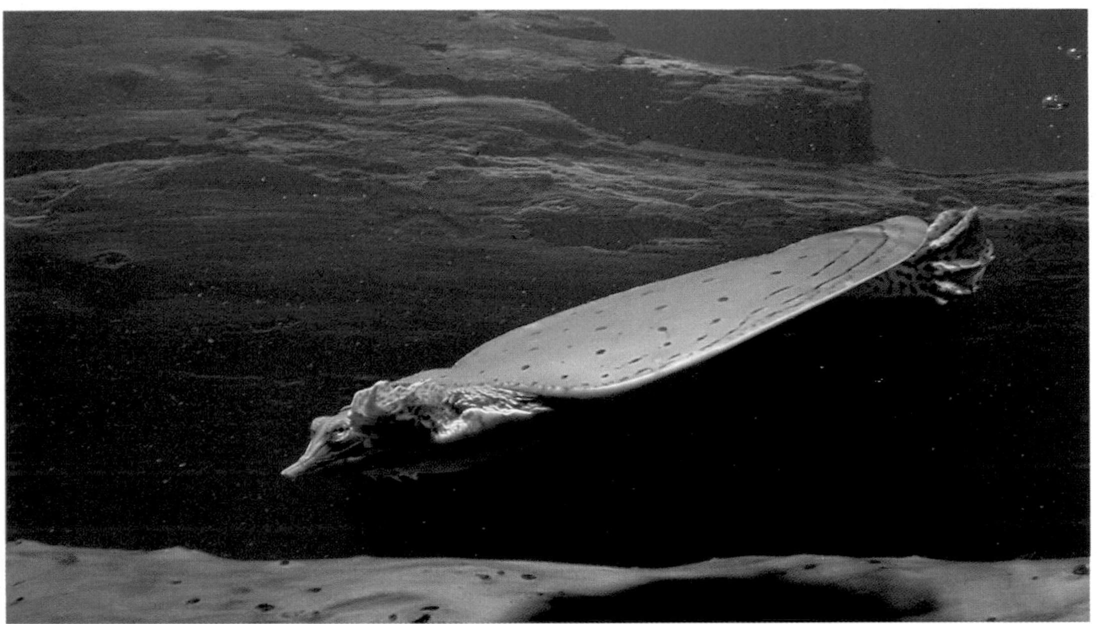

Ghost Crab (*Ocypoda ceratopthalma*)

TOP. This small crab is a shoreline scavenger which usually lurks in its burrow in the sand, its telescopic eyes scanning the beach for food. Under stress it blows bubbles from its mouth, partly as a defence reaction, partly to increase the oxygen content of the water retained in its body.

Eastern Spiny Softshell Turtle (*Trionyx spiniferus*)

BOTTOM. An inhabitant of the Gulf of Mexico, it has a bony shell covered with thick skin and numerous small knobs, and an elongated head ending in a proboscis, fleshy lips and horny jaws. It is a foot and a half long and eats any animal it can overcome. It only leaves the water to lay its eggs.

Nile Monitor (*Varanus niloticus*)

TOP. In spite of its name, it is found throughout Africa south of the Sahara. It grows to seven feet long and is an excellent swimmer, using its powerful tail as a rudder. Its diet includes snails, fish, frogs, toads and crocodile eggs.

Ganges River Dolphin (*Platanista gangetica*)

BOTTOM. Distinguished by its elongated snout, this is the most primitive of all dolphins and whales. Under the skin of its rectangular fins, it still has the five fingers of its land ancestors. It has tiny blind eyes without lenses and it navigates the muddy Ganges by sonar and sense of touch, using its snout to churn up the bottom to disturb fish and crabs. One of only four species of freshwater dolphins, it is the only mammal, apart from humans, that is known to swim on its side.

African Marabou Stork (*Leptoptilos crumeniferus*)

OPPOSITE. Photographed in Namibia's Etosha National Park, the fearsome African Marabou has a wingspan of up to 10 feet and the habits of a vulture. Its wedge-shaped beak is designed for cutting open the abdominal wall of dead animals. Its bare head is stripped of feathers so that any infectious bacteria that are picked up from the putrid flesh it eats will die in the heat of the sun. It also defecates on its own feet, a habit which, paradoxically, is also an antiseptic measure.

Northern Fur Seal (*Callorhinus ursinus*)

ABOVE. This is *the* fur seal of commerce and some 25,000 a year are killed in the name of fashion. It lives primarily on the Pribilof Islands in the northern Pacific, a territory the US acquired from Russia in 1868. The males of the species grow to seven feet long and weigh 600 lbs. During the breeding season, which starts in June, they establish territories and collect harems of up to 50 cows. In November and December, the Northern Fur Seals embark on their annual migration, swimming down America's Pacific coast. The females and young travel furthest, sometimes reaching San Francisco 3,000 miles to the south. They feed on squid, herring and other fish and are themselves preyed on by killer whales.

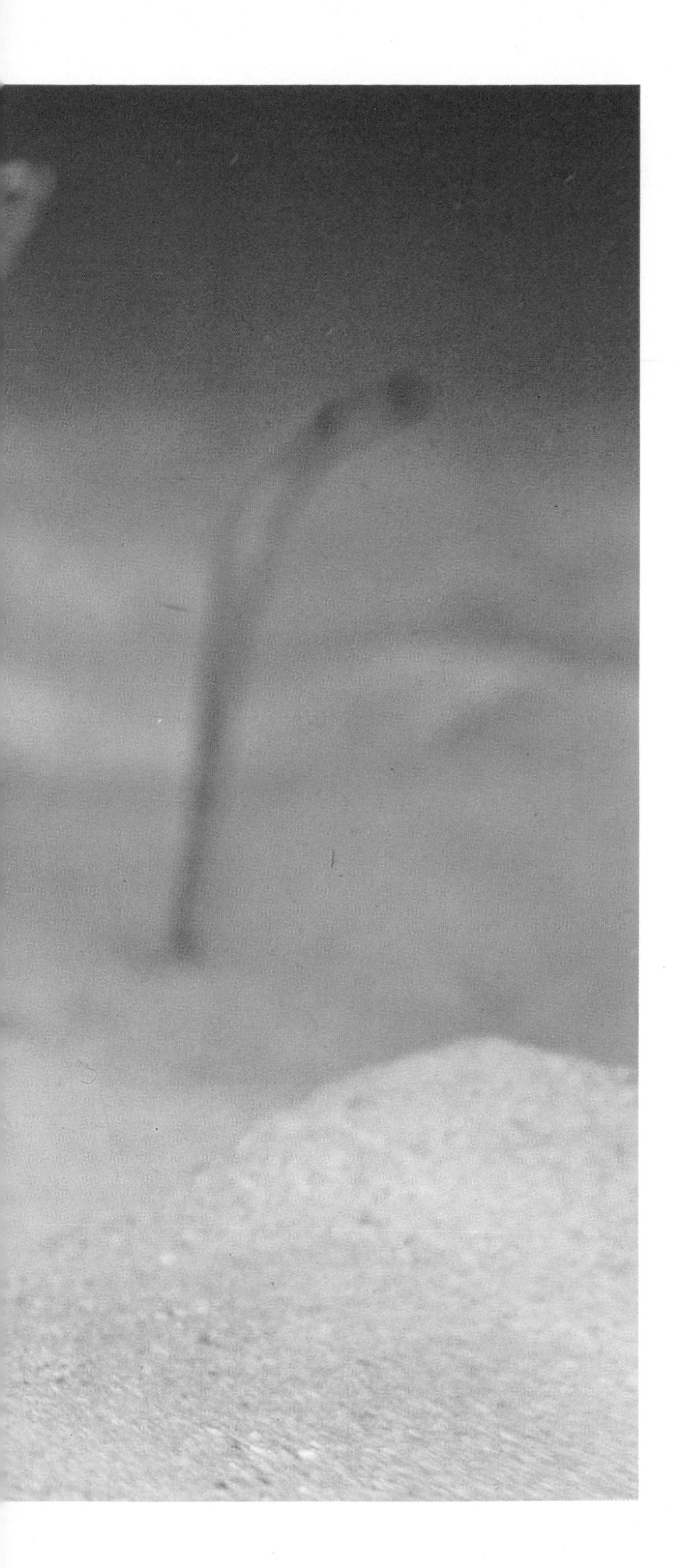

SEVENTEEN

GROUPS, MASSES & COLONIES

In contrast to the last section, these photographs show animals together. Social life occurs at all levels of the animal kingdom, from microscopic organisms to the largest mammals. Many creatures mass together for breeding purposes. Others form herds or flocks or schools because it minimises the chance of any one individual being singled out by a predator. Others form true colonies, where each individual is physically attached to its fellows. The numbers involved in these gatherings of animals often defy the imagination. Flamingo flocks may consist of hundreds of thousand of individuals, while swarms of insects or migrating birds can be counted in millions. We are not so different ourselves, after all, constructing massive nests which can contain, as in the case of Mexico City, over 10 million of our species.

Garden Eels (*Taenioconger hassi*)

These extraordinary fish grow a foot and a half long and live in tubes in the sand along with thousands of other eels, forming "gardens" that often cover 100 square yards. They sway to and fro in the water like weed, feeding on planktonic organisms carried to them by the current. They dig vertical burrows about 20 inches deep with powerful movements of their tails and retreat inside them if danger threatens.

Atlantic Sea Lion (*Otaria flavescens*)

This colony was photographed in December on cliffs south of Cape Bahias near Camarones in Patagonia. Sea lions are distinguished from seals by their lack of underfur, their size – the males of this species reach a length of eight feet – and their powerful, lion-like roar. They are playful creatures, especially the young, and are continuously in and out of the water since they do not like to get completely dry.

Starfish (Asteroidea)

It is difficult to equate these supple red creatures draped over a surf-washed rock on the coast of Peru with the hard, dried specimens found in natural history museums. The five symmetrical arms are semi-autonomous bodies; if one is torn off, the starfish grows a new one, while in some species the lost arm can itself regenerate a whole new animal. The starfish's body is ringed internally with the canals of a unique pumping system; sea water is sucked in through the animal's back and pumped along the arms and out through tube-like feet. These tube feet are also its means of locomotion – it expands and contracts them to "walk" along the seabed. They are tipped with suckers which the starfish uses to prise open the shells of clams and other molluscs. It then throws its stomach out through its mouth, envelops the soft part of its prey and gradually digests it.

Great Dusky Swift (*Cypseloides senex*)

ABOVE. Roosting on the rock face behind the Iguassu Falls in South America, these birds, also known as Temminck's Swift, spend their whole life near water. The ornithologist W. Meise, writing in *Grzimek's Animal Encyclopedia*, says of them: "In order to reach their nesting sites and resting places, the birds must fly through and across falling curtains of water, foam, and spray. It occasionally happens that a bird gets caught and swept away by a sudden torrent of falling water, but after a brief struggle it is usually able to free itself."

Red-Winged Blackbird (*Agelaius phoenicius*)

OPPOSITE. The tendency of modern agriculture towards large corn and wheat fields has led to a population explosion of this species and it is now probably the most numerous American bird south of the Canadian border. They migrate south in huge swarms after breeding in the spring and have become a pest in many areas. Males are black with scarlet epaulettes and mate with up to six females. The females are a uniform striped brown and build cup-shaped nests attached to water reeds in which they lay four to five eggs.

Chilean Flamingo (*Phoenicopterus chilensis*)

TOP. This breeding colony of flamingos, photographed 14,000 feet up in the Peruvian Andes, consists of pink-feathered adults and grey-brown juveniles. Flamingos live in huge flocks of up to several hundred thousand individuals on the edges of inland salt lakes or brackish coastal lagoons which provide ideal conditions for the crustaceans, algae and protozoa on which they feed.

Surgeon Fish (*Acanthurus achilles*)

BOTTOM. These fish are known as Surgeons because they have sharp bony protrusions like scalpels on either side of their tail. Although they can inflict deep and painful wounds, these weapons are purely defensive since Surgeons are strictly vegetarian. They have special chisel-like teeth for scraping algae off coral and thick stomach walls that enable them to digest the sand and coral particles they swallow as a result. Found in all tropical waters, they change colour to suit their mood. When they fight they adopt a "combat coloration" and after dark they have another colouring, known as their "nightgown". There are 100 Surgeon species, the largest growing two feet long.

Soldier Crab (*Mictyris longicarpus*)

TOP. Also known as Grenadier Crabs, they live along tidal estuaries and spend most of their time sorting through the mud looking for food. They often assemble in large groups like this one. These gatherings have been dubbed "sports clubs" because two or more of the crabs often scuttle along at high speed as if racing each other.

Pacific Walrus (*Odobenus rosmarus divergens*)

BOTTOM. Packed together and apparently fast asleep in a rocky bay of the Alaskan coast, huge herds like this are an increasingly rare phenomenon. The population of the Pacific Walrus was estimated at 200,000 in the mid-nineteenth century; today it is less than 45,000. Their tusks are considered a cash crop by the Eskimos, who also use every other part of the animals for food, clothing and shelter. But it is American and European hunters who are mostly responsible for the decimation of the species. Walruses are normally brown in colour, but turn a rich pink when they lie in the sun due to dilation of their surface blood vessels. (See also page 34).

Long-Tongued Fruit Bat (*Eonycteris* sp.)

These wall-to-wall bats in their cave on the island of Bali should be known as flower bats rather than fruit bats since they feed almost exclusively on petals, pollen and nectar. Only four to five inches long, they have a distinctively pointed face and a tongue that can be extended for some distance. The four known species of *Eonycteris* are found throughout south-east Asia from Burma to Indonesia.

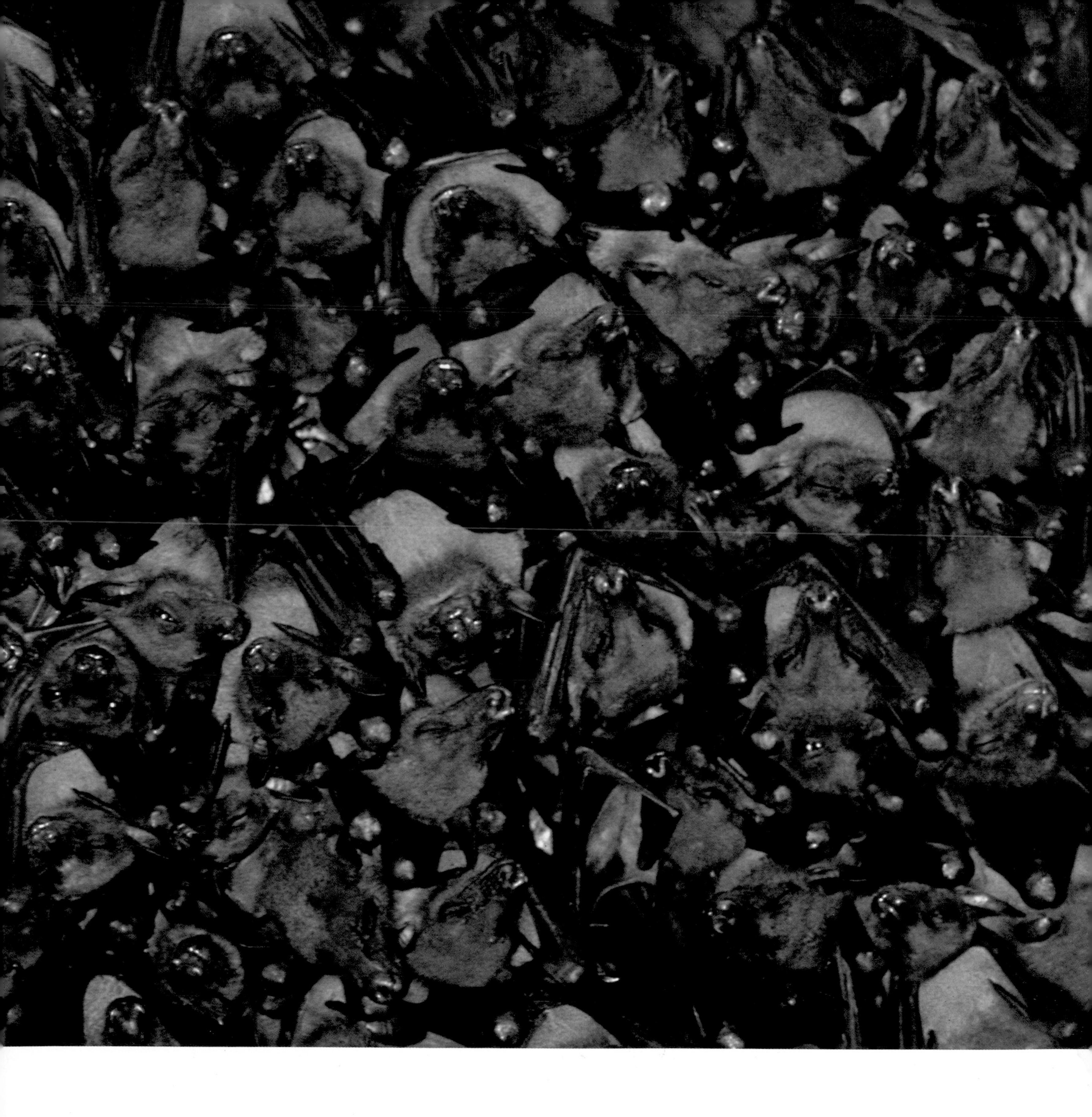

Hydrozoa

ABOVE. Seen under the microscope, this tree-like colony consists of minute marine creatures barely visible to the naked eye. They can reproduce both sexually and asexually. In the asexual form, a hydrozoan develops a bud on the wall of its body which grows until it becomes a separate individual. The new hydrozoan can either detach itself from its parent or remain attached to form colonies like this. There are about 2,700 species of hydrozoa, some 2,000 of which form colonies.

Bell Animalcules (*Epistylis rotans*)

OPPOSITE. This branching colony of microscopic, single-celled organisms leads a sedentary life. *Epistylis* belongs to the order Peritricha in which the hair-like cilia around the mouth are arranged in a left-hand spiral. The bell-shaped head is attached to a stalk muscle. These animals reproduce by simple longitudinal division, the offspring cell often going through a free-swimming stage before joining a colony.

Sea Squirts (Tunicata)

ABOVE & OPPOSITE. The Latin name for this group of animals refers to the gelatinous tunic surrounding their body. It is made of a substance called "tunicin", which is often known as "animal cellulose" because of its similarity to plant cellulose. Sea squirts are also distinguished by their unique circulatory system. They have no blood vessels, blood flowing instead through spaces in their tissue. The blood cells come in four colours – red, white, brown or green. The tubular heart pumps the blood in one direction through the body for about a hundred beats. It then stops altogether, before starting to pump in the opposite direction. This regular reversal of blood flow is found in no other animal. Some sea squirts are free-swimming throughout their life, but most species spend part or all their time as members of colonies. The cocoon-shaped colony above consists of thousands of individuals of the genus *Pyrosoma*. Most such conglomerations are only 5–10 inches long, but in one *Pyrosoma* species they can extend 30 feet or more. Each member of the colony has individual pores for inhaling and exhaling and a pair of light organs containing luminescent bacteria which glow in colours ranging from yellow to blue-green – hence the name *Pyrosoma*, meaning "fire-body". *Pegea socia* (opposite, top) is one of 20 species of so-called true salps. Each barrel-shaped individual is up to four inches long. They form chain-like colonies which, one naturalist observed, can be pulled along "like a line into a boat". This one appears to have a small crustacean caught in the chain. *Cyclosalpa pinnata* (opposite, bottom) forms a quite different colony. Each cone-shaped individual is a clone from the same parent, produced during the asexual stage of the animal's life cycle. The next stage occurs when an embryo grows inside each cone. When mature, they are squirted out into the water and swim off to start new colonies of their own.

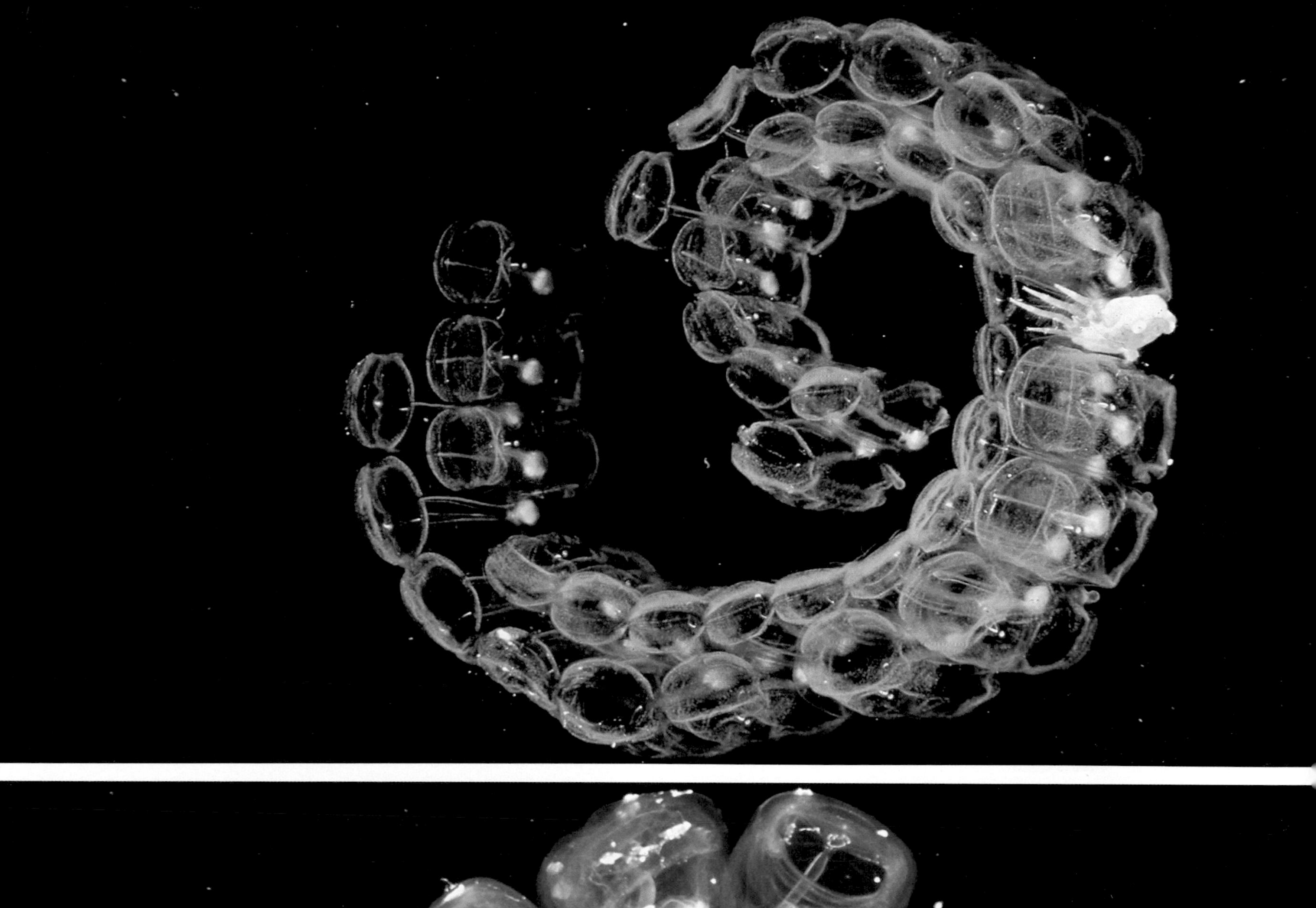

EIGHTEEN
UNDERWATER UFO's

The creatures commonly described as jellyfish come in two quite distinct forms during different phases of their lives, rather like caterpillars and moths. The familiar jellyfish form, with a more or less transparent, saucer-shaped umbrella and numerous tentacles hanging from it, is known as a medusa. When medusae reproduce, male and female release their sperm and eggs into the water to meet and fertilise. The resulting offspring are quite different-looking creatures, known as polyps. The polyp usually settles on the sea-floor and often sprouts like a tree with branching daughter polyps. After a time, these polyps produce buds which detach themselves and swim away to grow into the familiar medusa form. In some species, the entire polyp tree detaches from the seabed and becomes a free-swimming colonial jellyfish, with the various polyps that form the colony developing different shapes according to the function they perform.

Common Jellyfish (*Aurelia aurita*)

Distributed from the equator to near the poles, it grows up to 15 inches in diameter, with a fringe of long cilia around the edge of its disc. Its body is 98 percent water, making it only a little more dense that its environment. Food, oxygen and waste are moved around its body by a current of water which runs through internal canals. Light-sensitive organs enable it to sense the surface of the sea and balancing organs called statocysts help it to stay upright. Its body is transparent and almost colourless except for the horseshoe-shaped gonads which can be seen shining through. It feeds largely on planktonic organisms such as the larvae of crustacea, using its long cilia to wave them towards eight "food pits" on its upper surface.

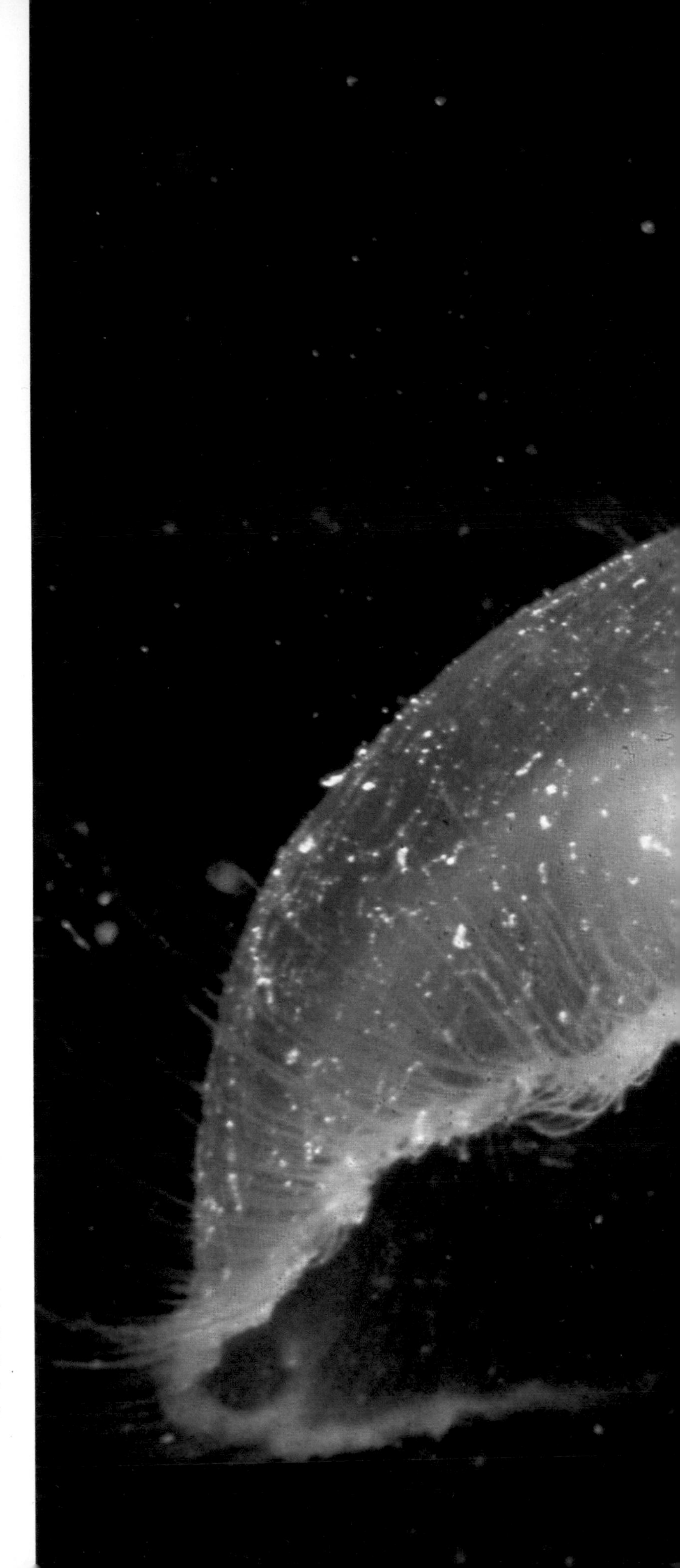

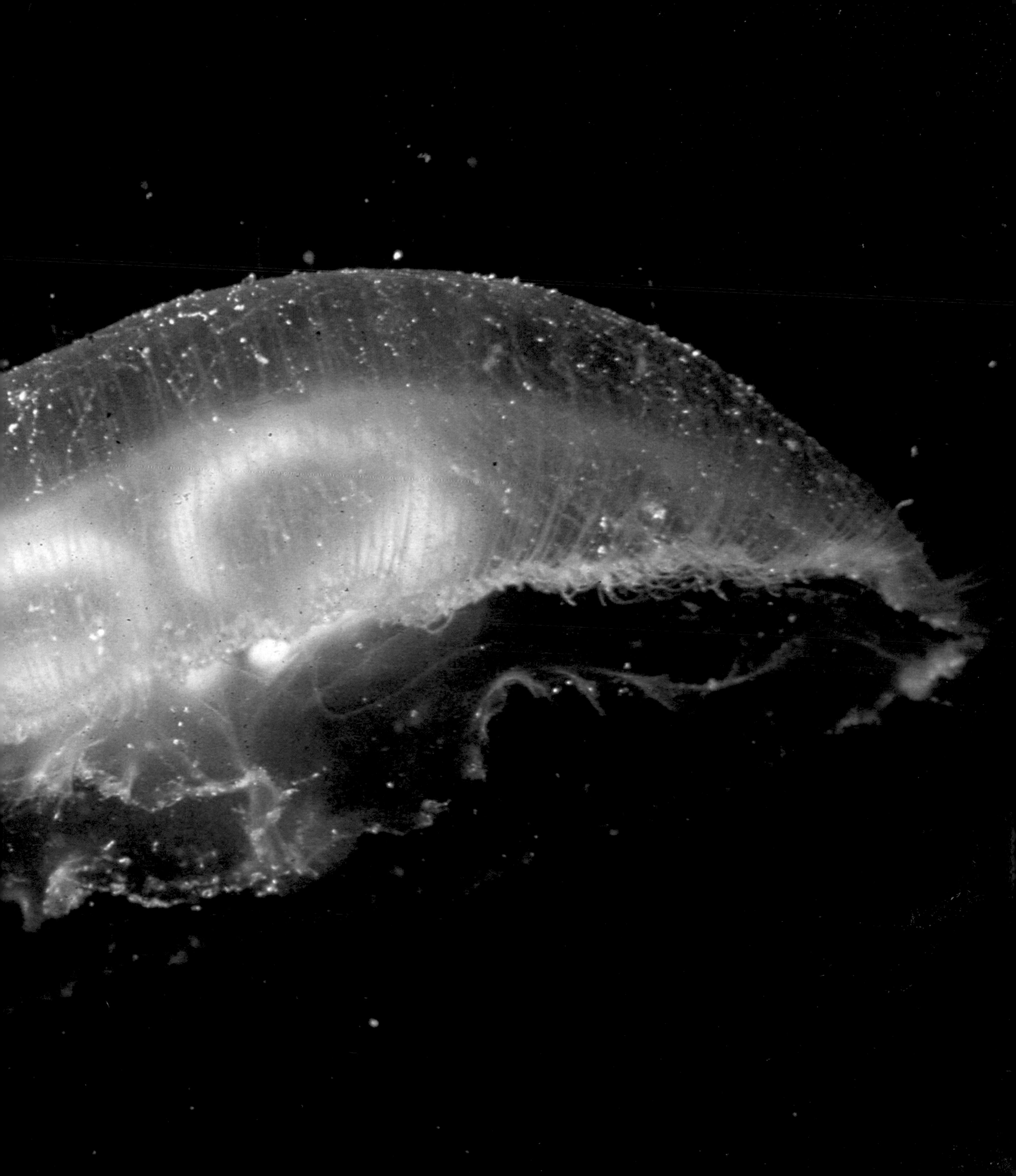

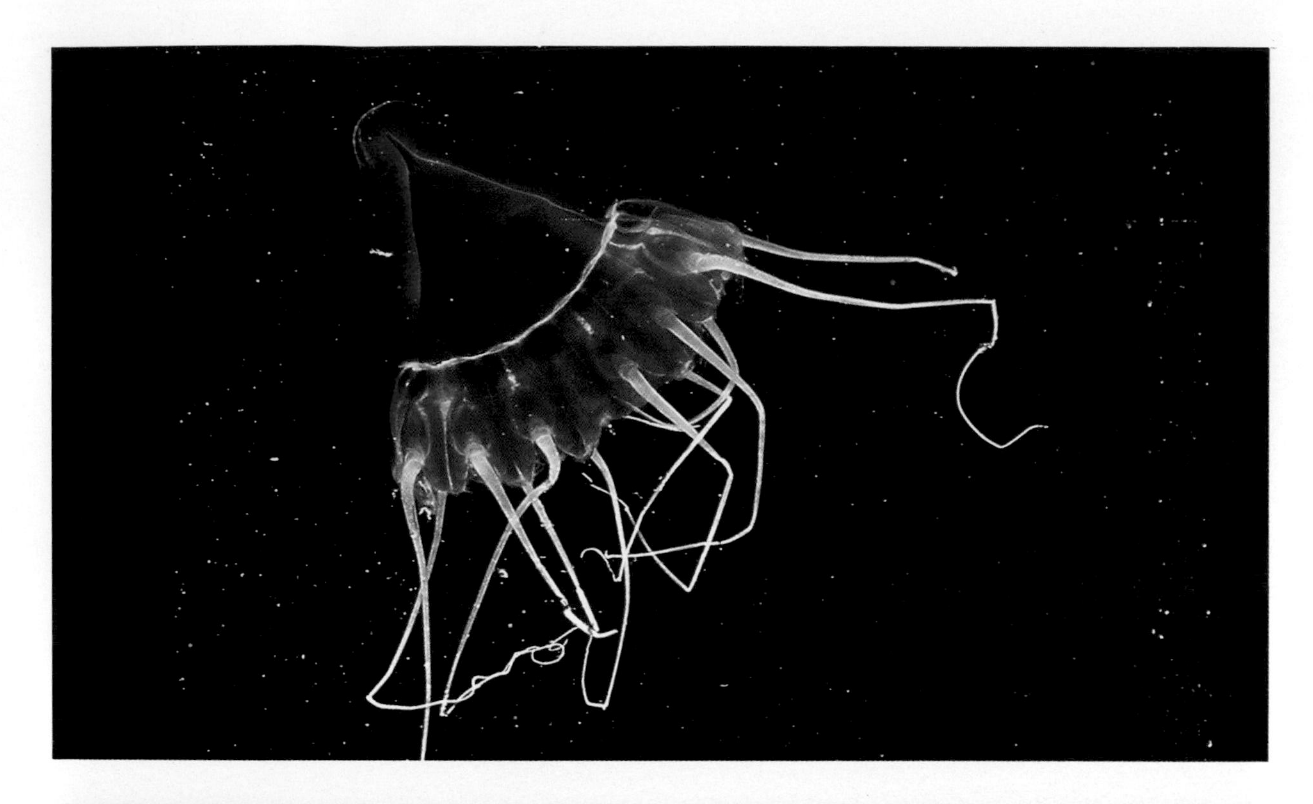

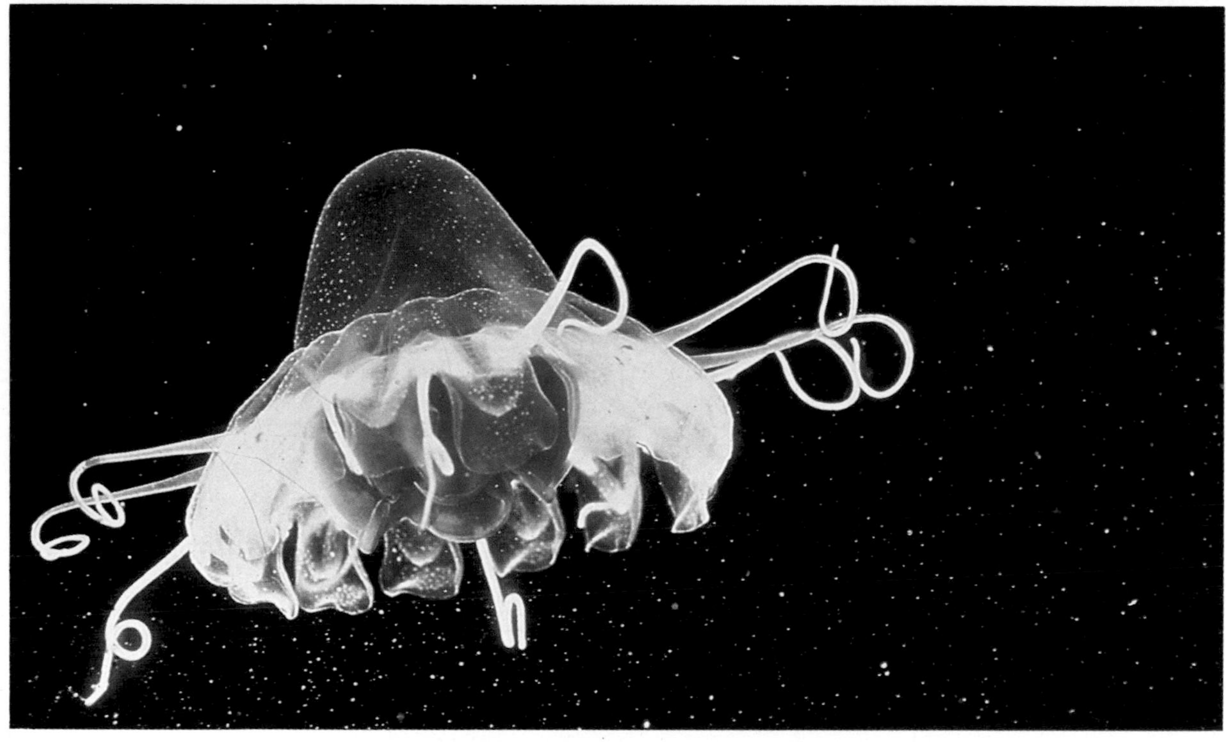

Deep-Sea Jellyfish (Coronata)

ABOVE & OPPOSITE. Although some species are occasionally found near the surface, most deep-sea jellyfish are found at considerable depths, down to 25,000 feet or more. The deeper they live, the more brightly coloured they tend to be. Most are very small, their "umbrellas" having a diameter on average of no more than two inches. They float like tiny alien space probes in the darkness. The ones on this page both belong to the genus *Periphylla*, the one opposite to the genus *Nausithoe*.

Comb Jellyfish (*Mnemiopsis maccradyi*)

This is one of about 80 species of comb jellyfish (Ctenophora) which are named for the eight rows of comb-like plates that run down the sides of their bodies. The beating of the combs gives these animals their propulsion and at the same time causes a rainbow of colours to move in waves along the plates, due to interference effects. Their bodies are so delicate that they are torn to pieces in rough seas, but they sometimes surface in huge swarms on nights of complete calm, turning the ocean into a carpet of colour. They have retractable tentacles equipped with adhesive cells that stick to the planktonic organisms and fish eggs they eat.

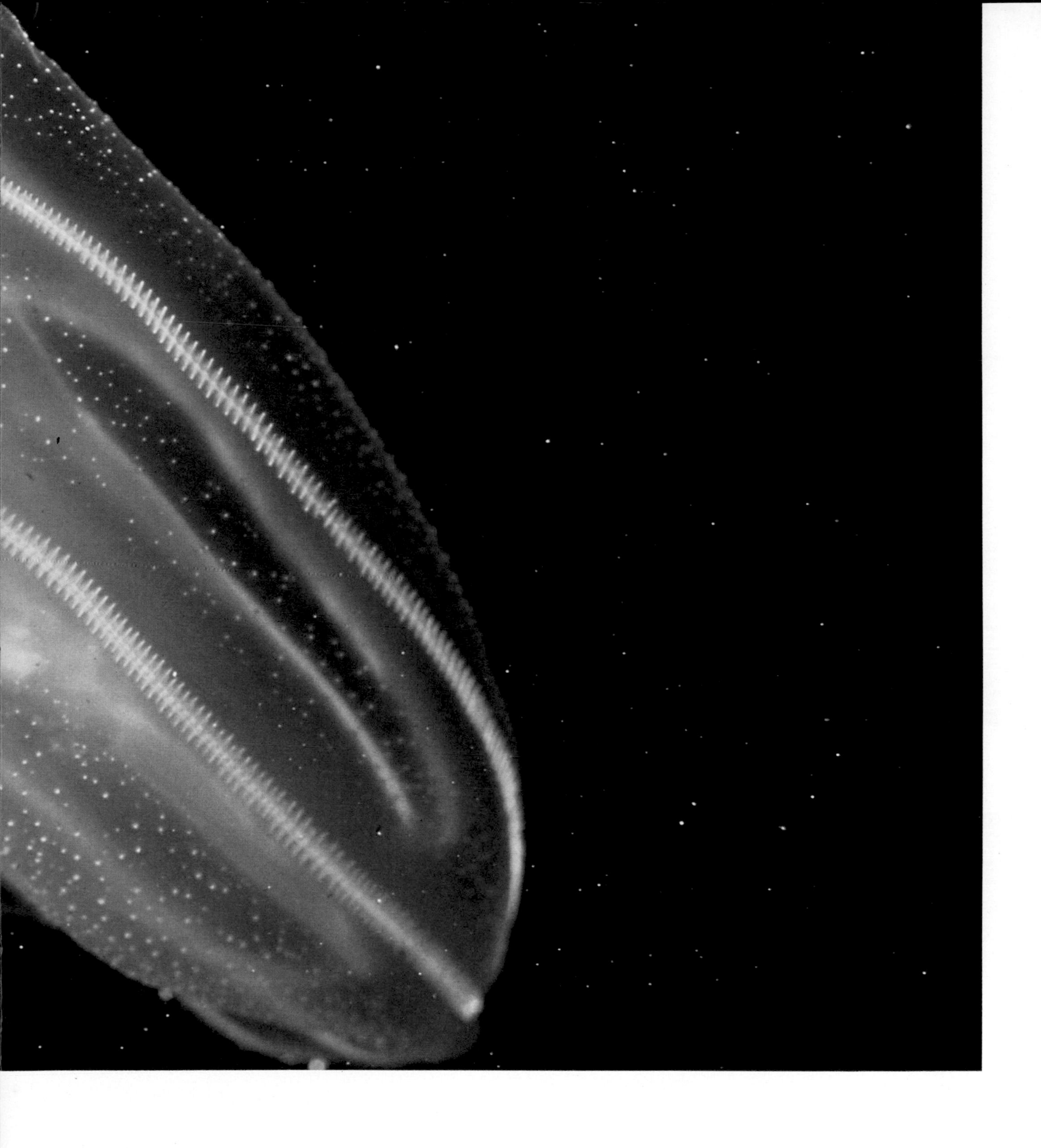

Siphonophore (*Physophora hydrostatica*)

ABOVE. Although it appears to be a single creature, this is a colony of several individuals who are differently shaped according to whether they perform feeding, defence, sexual or other functions for the colony as a whole. This species of siphonophore is closely related to the better-known Portuguese Man-Of-War. It is found mainly in the Atlantic but sometimes gets carried along by the Gulf Stream into the North Sea.

Calycopsis typa

OPPOSITE. Found in open waters of the north-west Atlantic, the thin-walled body of this colonial jellyfish has 16 internal canals which terminate in long tentacles whose bright tips are filled with stinging cells. The strangely brain-like fleshy clump at the centre of the body mostly consists of gonads which squirt sperm or eggs into the water for fertilisation.

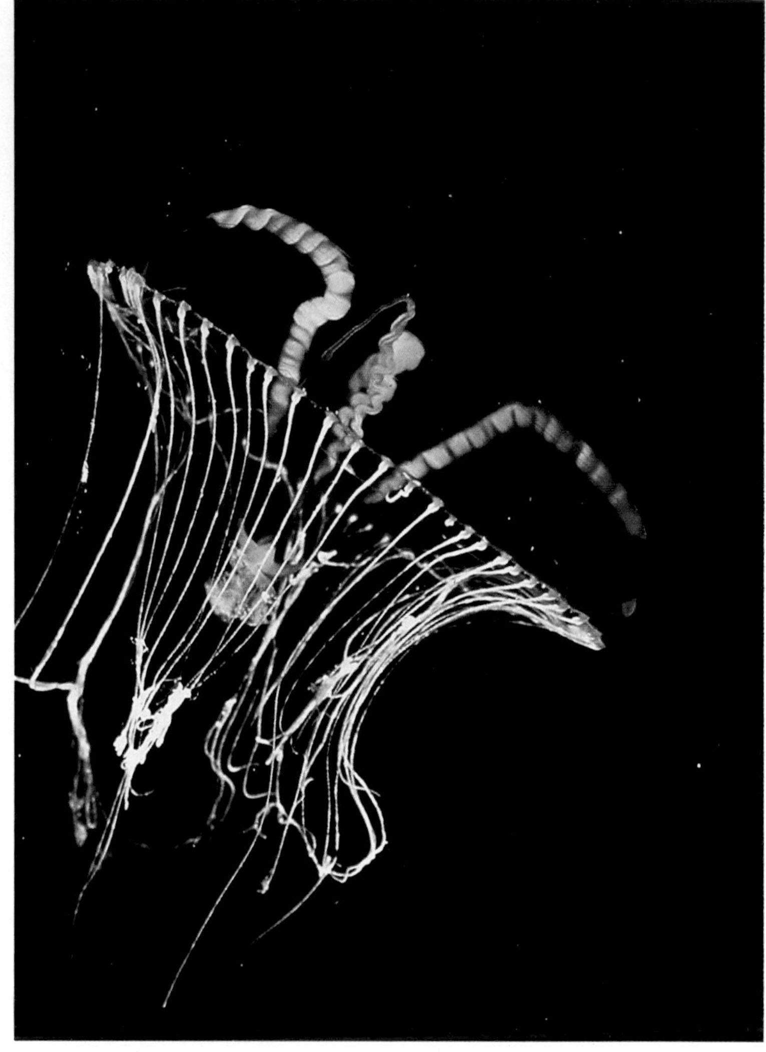

Lion's Mane Jellyfish (*Cyanea capillata*)

ABOVE, LEFT. Named for its yellow tentacles, thought to resemble a lion's mane, it inhabits the northern Atlantic and Pacific Oceans, the North Sea and the Baltic. Its umbrella is a foot in diameter and has eight clumps, each of 150 tentacles, hanging from it. When it captures prey, it spreads these poisonous tentacles like a net and gradually sinks to envelop its victim. It achieved literary fame when its poison was criminally employed in Conan Doyle's Sherlock Holmes story, *Adventures With The Lion's Mane*. Swarms of Lion's Mane Jellyfish over half a mile long have occasionally been sighted off the Norwegian coast. Better to meet one of these, however, than the creature's giant relative, the Arctic Lion's Mane. This is a monster amongst jellyfish, with an umbrella over six feet in diameter and tentacles which can be extended to an incredible 130 feet, forming a net over 500 yards square.

Tima flavilabris

ABOVE, RIGHT. A free-swimming jellyfish of the north-west Atlantic, its bell has a diameter of about four inches and is distinguished by the many tentacles trailing from it. The animal's polyps live in attached colonies. Its mouth is described as consisting of four frilled lips.

Porpita porpita

These are two views of one of the most highly developed colonial jellyfish. The view from above (top) shows the central disc which supports the colony when it comes to the surface. The disc is only about two inches in diameter and is surrounded by stinging tentacles, as seen in more detail in the view from below (bottom). Reproductive members of the colony, known as "sexual persons", are situated between the tentacles and the centrally placed mouth beneath the disc. *Porpita* frequents the surface of warm ocean waters throughout the world and is often found in huge drifting shoals.

Melon Comb Jellyfish (*Beroe porskallia*)

The melon combs of the genus *Beroe*, such as the pink, bell-shaped individual on the left, lack both the tentacles and adhesive cells of other comb jellyfish (see pages 216–217). They are little more than an enormous mouth whose interior is lined with poison glands for paralysing their prey. They feed exclusively on other comb jellyfish and have a voracious appetite. The animal in the background (*Eurhamphea* sp.) may well be intended as this melon comb's next meal.

PHOTO CREDITS

Front cover: Toni Angermayer. **Back cover:** Günter Ziesler.
Endpapers: butterfly wing photos by Dr Kjell Sandved, first published in *Smithsonian* magazine (January 1975). Page 2 Neville Coleman/Australasian Marine. 6: Firth & Firth/Bruce Coleman Ltd.

A SENSE OF PLACE. 10: Günter Ziesler. 12–13: Geoff Renner/Robert Harding Picture Library. 14–15: Jon Kenfield. 16–17: Alan Root/Bruce Coleman Ltd. 18-19: Delacour/Nature. 20–21: Roger Perry. 22–23: Laurence Gould/Seaphot: Planet Earth Pictures. 24–25: M.P.L. Fogden/Ecology Pictures. 26–27: Brian Hawkes/Robert Harding Picture Library. 28–29: Jon Kenfield.

HEADS. 30–31: Bill & Clare Leimbach/Robert Harding Picture Library. 32: Günter Ziesler. 33: Brian Hawkes/Robert Harding Picture Library. 34: Dr F. Sauer/ZEFA. 35: Frederic/Jacana. 36 (top): Bill Wood/Seaphot: Planet Earth Pictures. 36 (bottom): Dick Clarke/Seaphot: Planet Earth Pictures. 37: Brian Hawkes/Robert Harding Picture Library. 38: Ted Schiffman/Robert Harding Picture Library. 39: Bill & Clare Leimbach/Robert Harding Picture Library. 40: Philip Batson/Wildlife Picture Agency. 41: W. Roberz/ZEFA.

MICROWORLD. 42–51: all photos by John Walsh/Science Photo Library Ltd. 52–53: Kim Taylor/Bruce Coleman Ltd.

GALLERY OF THE GROTESQUE. 54–55: Heather Angel. 56: Lanceau/Nature. 57: Jane Burton/Bruce Coleman Ltd. 58–59: Jean-Philippe Varin/Jacana. 60: Jon Kenfield. 61: Vaughan Fleming. 62–63: Douglas Faulkner/Sally Faulkner.

ORCHIDS OF THE OCEANS. 64–65: Alex Kerstitch. 66 (top left): Dr T.E. Thompson/Science Photo Library Ltd. 66 (top right): Dr T.E. Thompson/Science Photo Library Ltd. 66 (bottom left): Dr T.E. Thompson/Science Photo Library Ltd. 66 (bottom right): Bill Wood/Seaphot: Planet Earth Pictures. 67: Peter David/Seaphot: Planet Earth Pictures. 68: Dr T.E. Thompson/Science Photo Library Ltd. 69 (top): Jon Kenfield, 69 (bottom): Dr T.E. Thompson/Science Photo Library Ltd.

PROPAGATING THE SPECIES. 70–71: Günter Ziesler. 72: Erwin A. Bauer. 73: Ferrero/Nature. 74: Densey Clyne/Mantis Wildlife Films. 75: Densey Clyne/Mantis Wildlife Films. 76: Peter David/Seaphot: Planet Earth Pictures. 77 (top): Hans Pfletschinger/Toni Angermayer. 77 (bottom): N.A. Callow/Robert Harding Picture Library. 78–79: Ferrero/Nature. 80: Günter Ziesler. 81: Champroux/Jacana. 82–83: Hans Reinhard/Toni Angermayer.

SURVIVAL STRATEGIES. 84–85: Alex Kerstitch. 86: Kim Taylor/Bruce Coleman Ltd. 87: Brian Hawkes/Robert Harding Picture Library. 88–89: Jim Frazier/Mantis Wildlife Films. 90–91: sequence of eight photos by Jim Frazier/Mantis Wildlife Films. 92 (top): Howard Hinton/Seaphot: Planet Earth Pictures. 92 (bottom): Peter Ward. 93: S. Bolwell/Agilis Pictures. 94–95: P.H. Ward/Natural Science Photos.

GREEN WORLD. 96–97: Günter Ziesler. 98 (top): Philippe Varin/Jacana. 98 (bottom): Wolfgang Bayer/Bruce Coleman Ltd. 99: Jean-Philippe Varin/Jacana. 100–101: Günter Ziesler/Toni Angermayer. 102: Günter Ziesler. 103: M.P.L. Fogden/Ecology Pictures. 104: Brian Rogers/Biofotos. 105: Hanif Raza/ZEFA.

BLUE FLIGHT. 106–107: L.R. Dawson/Bruce Coleman Ltd. 108 (top): Philippe Varin/Jacana. 108 (bottom): Richard Johnson/Seaphot: Planet Earth Pictures. 109: Heather Angel. 110–111: Soames Summerhays/Biofotos. 112: Stouffer Productions/Bruce Coleman Ltd. 113: Laboute/Jacana. 114–115: Douglas Allan.

EYES, SNOUTS & BEAKS. 116–117: Heather Angel. 118: Jim Frazier/Mantis Wildlife Films. 119: Kenneth Lucas/Seaphot: Planet Earth Pictures. 120–121: Dr James Porter. 122 (top): Warren Williams/Seaphot: Planet Earth Pictures. 122 (bottom): Jim Frazier/Mantis Wildlife Films. 123: Jim Frazier/Mantis Wildlife Films. 124: Rod Williams/Bruce Coleman Ltd. 125: Francisco Futil/Bruce Coleman Ltd. 126–127: Douglas Allan. 128: Douglas Faulkner/Sally Faulkner. 129 (top): Densey Clyne/Mantis Wildlife Films. 129 (bottom): Bill Wood/Robert Harding Picture Library. 130: Joseph Van Wormer/Bruce Coleman Ltd. 131: Dr Alan Beaumont. 132: M.P.L. Fogden/Ecology Pictures. 133: Günter Ziesler.

RIOT OF COLOUR. 134–135: Günter Ziesler. 136 (top): Heather Angel. 136 (bottom): Alex Kerstitch. 137 (top): Jon Kenfield. 137 (bottom): Bill Wood/Robert Harding Picture Library. 138: Günter Ziesler/Toni Angermayer. 139: Ian Took/Biofotos. 140 (top): Douglas Allan. 140 (bottom): N.A. Callow/Robert Harding Picture Library. 141 (top): Alex Kerstitch. 141 (bottom): Hans Pfletschinger/Toni Angermayer. 142 (top): Bill Wood/Robert Harding Picture Library. 142 (bottom): Soames Summerhays/Biofotos. 143: Peter Ward. 144: Günter Ziesler. 145: M.P.L. Fogden/Bruce Coleman Ltd.

PATTERN & TEXTURE. 146–147: R. König/Jacana. 148: A.J. Deane/Bruce Coleman Ltd. 149 (top): Jim Frazier/Mantis Wildlife Films. 149 (bottom): Densey Clyne/Mantis Wildlife Films. 150: Alex Kerstitch. 151 (top): Dr Alan Beaumont/Robert Harding Picture Library. 151 (bottom): Jon Kenfield. 152: Alex Kerstitch. 153: Douglas Faulkner/Sally Faulkner. 154: Queensland Museum/Natural Science Photos. 155: Soames Summerhays/Biofotos.

DISAPPEARING ANIMALS. 156: Jen & Des Bartlett/Bruce Coleman Ltd. 158: Peter Ward. 159: Heather Angel. 160: Peter Ward. 161: Douglas Faulkner/Sally Faulkner. 162: Densey Clyne/Mantis Wildlife Films. 163: David C. Rentz/Natural Science Photos. 164–165: Douglas Faulkner/Sally Faulkner.

CURIOUS CREATURES. 166–167: Jean-Philippe Varin/Jacana. 168 (top): Yves Kerban/Jacana. 168 (bottom): Graham Pizzey/Natural History Photographic Agency. 169: Peter Scoones/Seaphot: Planet Earth Pictures. 170: Jean-Philippe Varin/Jacana. 171 (top): Udo Hirsch/Bruce Coleman Ltd. 171 (bottom): Heather Angel. 172: Günter Ziesler/Toni Angermayer. 173: Günter Ziesler. 174–175: Alan Root/Bruce Coleman Ltd. 176 (top): Ferrero/Nature. 176 (bottom): David Cayleff/Wildlife Picture Agency. 177 (top): Kenneth Lucas/Seaphot: Planet Earth Pictures. 177 (bottom): Alain Compost/Bruce Coleman Ltd.

MONSTERS OF THE DEEP. 179: Peter David/Seaphot: Planet Earth Pictures. 180: Peter David/Seaphot: Planet Earth Pictures. 181: Peter David/Seaphot: Planet Earth Pictures. 182–185: Herve Chaumeton/Bassot/Nature. 187: Herve Chaumeton/Bassot/Nature.

ANIMALS ALONE. 188–189: Nat Fain/Natural Science Photos. 190: John M. Burnley/Bruce Coleman Ltd. 191 (top): Erwin & Peggy Bauer. 191 (bottom): Alain Compost/Bruce Coleman Ltd. 192: David Hughes/Bruce Coleman Ltd. 193: Geoff Renner/Robert Harding Picture Library. 194 (top): Heather Angel. 194 (bottom): Jack Dermid/Bruce Coleman Ltd. 195 (top): Dr Frieder Sauer/Bruce Coleman Ltd. 195 (bottom): F. X. Pelletier. 196: David Hughes/Bruce Coleman Ltd. 197: Eckart Pott/Photo-Center.

GROUPS, MASSES & COLONIES. 198–199: Allan Power/Bruce Coleman Ltd. 200: Kroener/Photo-Center. 201: Günter Ziesler/Toni Angermayer. 202: M.P.L. Fogden/Ecology Pictures. 203 Peter Ward. 204 (top): Günter Ziesler. 204 (bottom): Christian Petron/Seaphot: Planet Earth Pictures. 205 (top): Isobel Bennett/Natural Science Photos. 205 (bottom): P. Bading/ZEFA. 206–207: R. König/Jacana. 208: Kim Taylor/Bruce Coleman Ltd. 209: Dr Frieder Sauer/Bruce Coleman Ltd. 210: Douglas Faulkner/Sally Faulkner. 211 (top): L.P. Madin/Seaphot: Planet Earth Pictures. 211 (bottom): L.P. Madin/Seaphot: Planet Earth Pictures.

UNDERWATER UFO's. 212–213: Roy Manstan/Seaphot: Planet Earth Pictures. 214 (top): Peter David/Seaphot: Planet Earth Pictures. 214 (bottom): L.P. Madin/Seaphot: Planet Earth Pictures. 215: L.P. Madin/Seaphot: Planet Earth Pictures. 216–217: N. Swanberg/Seaphot: Planet Earth Pictures. 218: Heather Angel. 219: L.P. Madin/Seaphot: Planet Earth Pictures. 220 (top): L.P. Madin/Seaphot: Planet Earth Pictures. 220 (bottom): L.P. Madin/Seaphot: Planet Earth Pictures. 221 (top): Nat Fain/Natural Science Photos. 221 (bottom): L.P. Madin/Seaphot: Planet Earth Pictures. 222–223: L.P. Madin/Seaphot: Planet Earth Pictures.